PRAISE FOR SCOTT BRADFIELD'S *The History*

"A gripping tale of a haunted young mind and a penetrating, frightening symbolic look at the state of childhood in America. . . . The language [is] both spiritual and visceral, horrifying and humorous." —*The New York Times Book Review*

"Provides welcome evidence that...the poetry of America is not entirely dead. . . . Bradfield's sharp ear for the comedies and terrors of banality, his sense of the fear of disintegration in a universe of bewilderingly rapid change makes this a noteworthy debut." —*New York*

"An excellent debut, a display of talent in full self-possession." —*Times Literary Supplement*

"Casts an utterly irresistible spell." —Michiko Kakutani, *The New York Times*

"The triumph of mere existence is shown not only to be preordained but also seductively appealing." —*The Village Voice*

"It's Stephen King crossed with Joan Didion, perhaps with some of the hard gleam of Raymond Chandler thrown in as well. . . . " —*The Washington Post Book World*

"Painfully beautiful writing." —Mary Gaitskill

"A strange, brilliant novel." —*Elle*

"Chilling . . . intelligent, brilliantly written and often very funny . . . Bradfield's strange young protagonist constantly throws up poetic and alien insights." —*The Sunday Times*

"We hold our breath at Bradfield's calm way of writing about hair-raising events, his dead-pan brilliance." —*The Observer*

"Scott Bradfield's story of a boy growing up with his mother on the road is unsettling and bizarrely imaginative . . . a sharp satire of family values and the American dream . . . a dark, funny novel." —*The Daily Telegraph*

THE

HISTORY

OF

LUMINOUS

MOTION

THE
HISTORY OF
LUMINOUS
MOTION

SCOTT BRADFIELD

Picador USA
New York

Picador® is a U.S. registered trademark and is used by
St. Martin's Press under license from Pan Books Limited.

Grateful acknowledgment is made to the following for
permission to reprint previously published material:
Island Music Ltd.: Excerpt from "Many Rivers to Cross" by
Jimmy Cliff. Copyright © 1969 by Island Music Ltd. All
rights reserved. Used by permission.
Morley Music Co. and Cahn Music Company: Excerpt from
"It's Been a Long, Long Time" by Jule Styne and Sammy
Cahn. Copyright 1945 by Morley Music Co. Copyright
renewed 1973 by Morley Music Co. and Cahn Music
Company. International copyright secured. All rights
reserved. Used by permission.

Library of Congress Cataloging-in-Publication Data

Bradfield, Scott.
    The history of luminous motion / by Scott Bradfield.
        p.   cm.
    ISBN 0-312-14089-4
    1. Mentally ill children—California, Southern—
Fiction.   2. Mothers and sons—California, Southern—
Fiction.   3. Boys—California, Southern—Fiction.
I. Title.
PS3552.R214H56   1996
813'.54—dc20   95-46777
                                                CIP

First published in the United States of America by
Alfred A. Knopf, Inc.

First Picador USA Edition: April 1996
10   9   8   7   6   5   4   3   2   1

For Felicia

This is the long lulled pause
Before history happens . . .

—TOM PAULIN

THE

HISTORY

OF

LUMINOUS

MOTION

MOTION

1

Mom was a world all her own, filled with secret thoughts and motions nobody else could see. With Mom I easily forgot Dad, who became little more than a premonition, a strange weighted tendency rather than a man, as if this was Mom's final retribution, making Dad the future. Mom was always now. Mom was that movement that never ceased. Mom lived in the world with me and nobody else, and every few days or so it seemed she was driving me to more strange new places in our untuned and ominously clattering beige Rambler. It wasn't just motion, either. Mom possessed a certain geographical weight and mass; her motion was itself a place, a voice, a state of respose. No matter where we went we seemed to be where we had been before. We were more than a family, Mom and I. We were a quality of landscape. We were the map's name rather than some encoded or strategic position on it. We were like an MX missile, always moving but always already exactly where we were supposed to be. There were many times when I thought of Mom and me as a sort of weapon.

"Do you love your mother?" one of Mom's men asked me. We were sitting at Sambo's, and I was drinking hot chocolate. Mom had gone to the ladies' room to freshen up.

It seemed to me a spurious question. There was something sedentary and covert about it, like the bad foundation of some prospective home. I had, as always, one of my school texts open in my lap. It was entitled *Our Biological Wonderland: 5th Edition*, and I was contemplating the glossary to Chapter Three. I liked the word "Chemotropism: Movement or growth of an organism, esp. a plant, in response to chemical stimuli." Chemotropic, I thought. Chemotropismal.

"Your mother is a very nice person," the man continued. He didn't like the silence sitting between us there at the table. I myself didn't mind. He smoked an endless succession of Marlboros, which he crushed out in his coffee saucer rather than the Sambo's glass ashtray resting conveniently beside his elbow. Nervously he was always glancing over his shoulder to see if Mom was back yet. I didn't tell him Mom could spend ages in the ladies' room; the ladies' room was one of Mom's special places. No matter where we were living or where we were traveling, Mom found a sort of uniform and patient atmosphere in the ladies' rooms where she went to make herself beautiful. Sometimes, when I accompanied her there like a privileged and confidential adviser, we would sit in front of the mirror for hours while she tried on different lipsticks and eye shadows, mascaras and blushes. Mom found silence in the ladies' room, and in the beauty of her own face. It was like the silence that sat at the tables between me and Mom's men, only by Mom and me it was more appreciated, and thus more profound.

"I love my mom," I said, holding the book open in my lap. Mom's man wasn't looking at me, though. He seemed to be thinking about something. It was as if the silence had actually moved into him too, something he had inherited from the still circulating memory of Mom's skin and Mom's scent. I looked into my book again, and we sat together drinking our coffee and hot chocolate, awaiting that elimination of our secret privacy which Mom carried around with her like a brilliant torch, or a large packet of money. Sometimes I felt as if I were a million years old that summer, and that Mom and I would continue traveling like that forever and ever, always together and never apart. I remember it as the summer of my millionth year, and I suspect I will always remember that summer very well.

Those were nights when we moved quickly, the nights when Mom found her men. Usually I would lie in the backseat of our car and read my faded textbooks, acquired from the moldering dime bargain boxes of surfeited and dusty used-book stores. I would read by means of the diffuse light of streetlamps, or the fluid and Dopplering light of passing automobiles. Sometimes I had to pause in the middle of paragraphs and sentences in order to await this sentient light. In those days I thought light was layered and textured like leaves in a tree. It moved and ruffled through the car. It felt gentle and imminent like snow. Eventually I would fall asleep, the light moving across and around me on some dark anonymous street, and I would hear the car door open and slam and Mom starting the ignition, and then we would be moving again, moving together into the light of cities and stars, Mom pulling her coat over me and whispering, "We'll have our own house someday, baby. Our own bedrooms, kitchen and TV, our own walls and ceilings and doors. We'll have a brand-new station wagon with a nice soft mattress in back so you can lie down and take a nap any time you want. We'll have a big yard and garden. We might even have a second house. In the mountains somewhere."

In the mornings I would awake in different cities, underneath different stars. Only they were the same cities, too, in a way. They were still the same stars.

Mom kept the credit cards in a plastic card file in the glove compartment, even the very old cards that we never used anymore. The file box also contained a few jeweled rings and gold bands which we sold sometimes at central city pawnshops, and a few random business cards with phone numbers and street maps urgently scrawled on their backs. These were the maps of Mom's men, and sometimes I preferred looking at them rather than at my own textbooks. These were names of things, people and places that possessed color, suspense and uniformity, like a globe of the world with textured mountain ranges on it. Lompoc, Burlingame, Half Moon Bay, Buellton, Stockton, Sacramento,

Davis, San Luis Obispo. Real Estate, Plumbing, Fire Theft Auto, 24 Hour Bail, Good Used Cars, Cala Foods and Day-brite Cleaners. Mom's men were accumulations of words, like nails in a piece of wood. When I closed the plastic file again the lid's plastic clamp clacked hollowly. "That's Mom's Domesday Book you've got there," Mom said. "Her Dead Sea Scrolls, her *tabula fabula*. That's Mom's articulate past, borrowed and bought and certainly very blue. If they ever catch up with your old mom, you take that file box and toss it in the river—that is, if you can find a river. Head for the hills, and I'll get back to you in five to ten, though I'm afraid that's just a rough estimate. I've stopped keeping track of the felonies. I think that's the compensation that comes with age—not wisdom. You're allowed to stop keeping track of the felonies." Mom was wearing bright red lipstick, tight faded Levi's and a yellow blouse. She drank from a can of Budweiser braced between her knees. I didn't think Mom was old at all. I thought she was exceptionally young and beautiful.

Outside our dusty car windows lay the flat beating red plains of the San Fernando Valley. Dull gray metal water towers, red-and-white-striped radio transmitters, cows. "Emily Dickinson said she could find the entire universe in her backyard," Mom told me. "This, you see, is our back-yard." Mom gestured at the orange groves and dilapidated, sunstruck fresh-fruit stands and fast-food restaurants aisling us along Highway 101. The freeway asphalt was cracked and pale, littered with refuse and the ruptured shells of over-heated retread tires. Then Mom would light her cigarette with the dashboard lighter. I liked the way the lighter heated there silently for a while like some percolating threat and then, with a broken clinking sound, came suddenly unsprung. Mom's waiting hand would catch it—otherwise it would project itself onto the vinyl seat and add more charred streaks to the ones it had already made. There was even a telltale oval smudge against the inside thigh of Mom's faded Levi's. "Now, keep your eyes out for the Gilroy off-ramp," Mom said. "It's along here somewhere. We'll have a McDonaldburger and then I know this bar where maybe I'll get lucky. Maybe we'll both get lucky." And of course we always did.

2

Because I always identified Mom according to her customary and implicit movement, whenever that movement ceased or diminished it seemed to me as if Mom's meaning had lapsed too. It was her wordlessness I recognized first, that pulse and breath of her steady and unflagging voice. It was a soundlessness filled with noise, a meaninglessness filled with words. It was like that intensification of language where language is itself obliterated, as if someone had typed a thousand sentences across the same line of gleaming white bond until nothing remained but a black mottled streak of carbon.

"This is Pedro," she told me that long ceremonious day in San Luis Obispo. We had been staying the week at a TraveLodge on Los Osos Boulevard, thanks to the uncomprehending beneficence of Randall T. Philburn, a ranch supplies salesman Mom had met in a King City Bingo Parlor the week before. Randall had carried Diner's Club and American Express. He had shown me a trick with two pieces of string. The next time I saw him, I was supposed to have memorized the names and chronologies of all our presidents.

"And this, Pedro, this is the only important man in my life," Mom said. "My unillustrious and laconic son, Phillip."

So that was how it began. She told me his name was Pedro, as if all her men had names. Pedro. As if a man's name were something to be uttered and not a bit of embossed plastic to be stored in a grimy beige plastic file box in our Rambler's rattly glove compartment. Pedro. As if I were supposed to remember. As if a man's name was something you said with your mouth so that another's ears might hear.

It was no simpler than that, that first staggering cessation of Mom's body and her voice. Barely an utterance and more than a name. Pedro. And it wasn't even his name, really.

"How you doing, sport?" Pedro asked, teaching me a firm handshake. His real name was Bernie Robertson, and Bernie possessed a round florid face (particularly after his second or third Budweiser), a hardware store in Shell Beach, a slight paunch, and a two-bedroom house in the Lakewood district of San Luis Obispo, where I was allowed the dubious privacy of my own room. It was only a week after our first, formal introduction that Bernie helped us transfer our few things from the TraveLodge into his home where Pedro's real, unvoiced name was everywhere. It was on the mail and on the automobile registration and on the towels and on the hearth rug, it was on the mortgage and the deed. It was even burned into a crosscut oak placard which hung from Pedro's front porch: THE ROBERTSONS. It was a name which, unless we were very careful, might very soon attach itself to both Mom and me.

"My house is your house," Pedro liked to say, sitting on the sofa with his arm around Mom, his can of Bud balanced on his right knee. Pedro's house contained stuffed Victorian love seats, knickknack shelves, porcelain statues of Restoration ladies and gentlemen engaged in rondels and courtly kisses, untried issues of *Reader's Digest* and *The Saturday Evening Post*, lace doilies and even antimacassars. Mom lay with Pedro on one sofa, her head in his lap, his arm across her breast. I sat alone on the love seat with my textbook. It was entitled *Science and Our World Around Us*, and contained a color photograph of *E. coli*. Most human beings and animals contained this bacterium in their intestines, the photo caption said, and though generally benign, it could cause infant diarrhea and food poisoning. Mom and Pedro seemed very happy and warm there in front of the fire. The television was on, generating its soft noise. One slice of dry pizza remained in the oily cardboard container beside the blazing brick fireplace where Heidi, Pedro's smug and disaffected gray cat, paused occasionally in its rounds to lick at it. Sometimes I just read the dictionary. Auto-da-fé, autodidact, autoecious, autogamy, autoimmune. Words in a dictionary have a rhythm to them, a dry easy meaning I can assemble in my head like songs, or caress like pieces of sculpted wood. Autoecious, I thought. Autogamy. Autoimmune.

"Is there anything you'd like to watch, Phillip?" Mom might ask. "Pedro and I just watched a program *we* wanted to watch."

I disregarded Mom's offer, drifting in the currents of words and pictures issuing from the privacy of my own books. The television remained tuned to whatever mundane channel Pedro and Mom had selected. With the conclusion of that sad summer, I was casually enrolled in school.

Needless to say, my first experience of public education was at once harrowing and nondescript. There was something nightmarish about the actual absence of terror in that place, which always struck me as a sort of systematic exercise in vaguely hollow and uneventful routine. There were other boys and girls there of my own age who I was encouraged to get to know. When I didn't speak at the daily Show and Tell, my reticence was attributed to shyness and not intimate revulsion. Stories and fairy tales were read out loud to us, and we read to ourselves tedious true-life stories from the flimsy plastic pamphlets of the SRA Reading Program. (I was assigned to the intermediate level Red, due to my own deliberate stumbling over consonantal clusters and mixed vowels. I was determined none of these strangers would know me.) We were there for seven or eight hours each day. Games, talk, asinine books, endless recesses, stupid unsatisfiable pets in cages lined with their own urine and sawdust, colored paper and paste and scissors which we were told to hold *in* to our bodies when we passed them back to our Art Supplies Monitor. (Every one of us was designated by some such atrocious insignum, like cabinet officials in

12

some tawdry, self-important South American nation. I, for example, was Chalk Board Clearance Superviser.) It was interminable day after day of vacuous and unremitting childhood, unrelieved by any useful information whatsoever. The world had closed itself around me, and threatened to teach me only what it wanted me to know.

"Your mom's a real special woman, one first-class lovely lady," Pedro liked to assure me. Every afternoon we were usually alone together for an hour or so after school, because Mom had taken a part-time job at the local Lucky Food Store, boxing groceries. "You're a very fortunate young man to have a mother who loves you so much." He was never really looking at me when he spoke, but rather pulling the pop tops off beer cans, fiddling with the TV's horizontal and vertical controls, or building something useful in the backyard. He didn't really speak so much as erupt with aphorisms. "Everybody needs to settle down someday," he might say, or, "Sometimes a woman needs somebody who can take care of her, too. Even mothers need a little love and support sometimes." Then he would clamp something to the steel vise, or shave the spine of some unvarnished plywood door. On sunny days we hauled his tools and machinery out to the splintering pine workbench in the yard, and in those dull equivocal months of Mom's immobility the days seemed relentlessly sunny. I would stand and watch from a distance—not for self-protection, but simply because I didn't want to get too involved. Pedro loved to build things out there: a trellis, picnic table and chairs, cement patio,

13

brick fireplace. If there were world enough and time I'm sure Pedro would have built airport runways out there, enormous ivory mausoleums, pyramids and skyscrapers and spaceships and planets. With the hacksaw which I always watched him replace so carefully in his oiled and immaculate toolbox. With the pliers. With the sharp steel file. With the ball peen hammer. With all those solid and patently useful tools he kept filed in the large glimmering steel toolbox and stored underneath the same bed in which he and Mom slept together each night. It seemed the appropriate place to keep them, I thought. They massed underneath there like weather; you could feel the pressure of them in other rooms and houses. With these tools Pedro had built things in Mom's mind too, working late at night while she slept. There was a literal or figurative truth in that image for me, and, during those horribly persistent days of domesticity, I didn't care which was more correct. The literal or the figurative.

Mom began doing strange things after we moved into Pedro's house. She whistled sometimes, or sewed curtains. She darned socks. She even embroidered. I remember sitting beside her on the sofa and watching her hands fumble with the lacy cloth and a sharp, gleaming needle. That needle was the only part of the entire process which seemed to make any sense to me. The needle was abrupt and binding. It carried with it its own sharp logic. Then one day the new curtains were hung in the living room, and Mom put her hands on her hips and smiled. I suppose I was expected to smile too, but I didn't smile. I looked at the thin curtains, though. They seemed to me just perfect for Pedro's thin house.

"Are you happy?" she would ask me during our private talks late at night, for Pedro always went to bed and awoke early.

"I guess."

"Are you making friends at school?"

"I guess."

"Do they ever invite you to their houses? Do they have nice families who make you feel welcome?"

"Sometimes," I said, on very shaky ground now. I didn't know what Mom expected from me. "Sometimes, you know, well. We don't go anywhere. We just sort of sit around, you know."

"Do they have nice yards and gardens?"

"Some of them, I guess."

"Do they have nice rooms filled with nice toys?"

"Sure, some of them."

Mom smiled brightly, but without looking at me. She was crushing out her Marlboro and gazing off into the bright rooms and gardens of my imaginary friends.

"It's better for you this way," she said. "You deserve a normal upbringing, some firm and certifiable life. It's the only time life is certifiable, baby. When you're a child. When you grow up it doesn't make any sense, whatever way you look at it. Would you like to bring one of your new friends home for dinner some night?"

"I don't think so."

"Do you like your new home? Do you like Pedro?"

I thought for a moment. I felt a hot, blazing fire swelling up in my heart, my face, my vision. My throat constricted. I felt suddenly dizzy and blurred. Hoarsely I answered, "He's a nice man, I guess."

"You're right, baby," Mom said. "He is a very nice man."

One unforgivable day Mom even took me to Penney's for what she referred to as my "school clothes," and for one wild catastrophic moment pulled me to a halt beside the racks of Cub Scout uniforms and supplies. Compasses and safety knives and handkerchief rings and merit badges and handbooks and tents. Finally she bought me white wool socks, cotton underwear and a map of the solar system, which she posted on the wall of my room in order to provide what she called a "vigorous bit of brain food," just as proud moms everywhere hang glittering mobiles above the cribs of their dully gazing babies. I always kept expecting things would get better. Instead they just got worse and worse.

There was talk of a birthday party in November, and throughout that entire summer I paced and worried in a monstrous imminence of cakes, candles, other children in foil hats, door prizes and gifts with bright wrapping and scissor-scored, frilly ribbons. We would play party games at this "birthday party." Mom would award little prizes, and be careful no one child was overlooked. I would close my eyes and make wishes. I would greet all my beribboned friends at the front door with an ingratiating look on my face. Ultimately, while Pedro cheerily drank his beers and reacquired his customary flushed smile, I would be ceremoniously required to open presents. New shirts, model planes, transistor radios, "young adult" books, record albums, perhaps even my own portable record or cassette player. Board games, desk lamps, magazine subscriptions, boats and T-shirts and socks. Things and more things, accumulating in my lap, pulling the weight down out of my abdomen, pulling both Mom and me closer to the hard ground, deeper into the intractable earth. Nothing but weight and gravity and mass, immovable mass. And that look of motionlessness in Mom's once beautiful eyes. "For your next birthday, we'll have a party in the park," she would tell me, just as I thought the ordeal was over, still wiping the slightly hysterical tears from my eyes. "You can invite even more friends. You'll receive even more presents."

At this I would awake with a sudden start in my sweaty bed, entangled in my twisted blankets, surrounded by the concrete moonlight, enveloped by the whirling dust. The solar map confronted me then like a graceless benediction, filled with cartoon colors and impossibly tidy convergences. Moons and planets and suns, imprisoned by gravity and centrifuge and chemical weight. Perihelion and apogee. Jupiter and Mars. I would have gladly disappeared into any of them. I would have boiled on Mercury, exploded with my own freezing expanding breath on Pluto. I felt all the movement coming to a stop inside me, like the gestating atmospheres of nascent planets. Someday Jupiter would be like that, a ball of impacted dirt, senseless rigid cities, malign

17

children assembled around some ominous birthday cake with their noisemakers and party hats. I was growing more solid and permanent every day. Perhaps people would even start calling me by a nickname. Buster, or Chipper, or Mac. I could easily imagine Pedro learning to call me Mac. I could even see the word as it was thickly articulated by his fleshy lips, as if he were extruding a soft rubber ball on the tip of his plump pink tongue. How you doing, Mac? How about we go out to a ball game, Mac?

I couldn't return to sleep. I tossed and turned. Before I suffered a real birthday party I would kill myself; I vowed they would all repent their relentless cruelties, and with a certain relish I selfishly imagined my own obituary and funeral. The day would be rainy and dark as they lowered my forlorn, tiny casket into the deep, sculpted earth. Mom would cry and cry, but there would be nobody to hold her like I could hold her. Mom would know then. She would know the horror and loneliness she had subjected me to. Pedro would stand firmly beside her, but there was nothing he could do to stop her crying. Convulsing, weeping, begging me to come back. Uranus, Neptune, Pluto, Planet X. As I diminished in Mom's universe, she could only stand helplessly by and watch me go. If I couldn't live in Mom's universe, then I would teach her. I would find a universe of my own. Those were nights when I actually and sincerely hated my mom. I may never forgive myself for it, but I really hated her then.

———

On rare occasions their bedroom door was locked at night, but usually they left it wide open, encouraging, I guess, some idyllic familial confidence and integrity. On nights I couldn't sleep I might go in there and look at them embraced by their bleached and complicit white sheets. Mom always slept on her right side, near the verge of their king-sized Stayrest mattress. Pedro grunted and snuffled in his sleep like a pig. His belly looked even bigger when he lay on his back, his mouth open, his splotchy face expressionlessly stupid. When I looked at Pedro sleeping I felt something vegetable and hard growing inside me. It whispered with tangled roots and burrs and weaving, fibrous fingers. It moved only at night. It was trying to tell me something about myself nobody had ever told me before. It reached through everything. It was almost here.

3

It was just a phase she was going through, I had convinced myself. Like menstruation or bad luck. Whenever Mom became maudlin and self-involved, I would just lay my head in her lap, wrap my arms around her and listen patiently, without offering her a word of reproach. "You deserve a better life than I ever gave you," Mom might whisper, holding an icy drink beside my ear, gazing aimlessly at her own reflection in the warped vanity mirror. "You deserve a home, baby. You deserve people you can count on, a place you can come to." I wouldn't say anything to Mom at these times. To say anything would only validate Mom's delusory self-recriminations. I was always certain we would start moving again at any moment. Mom was just resting; Mom was just recharging her batteries. Soon, without any fanfare, Mom would be Mom again.

So there I was, immured within Pedro's musty sanctum, my own fault, really. I had never read the signs correctly; I had not anticipated every swerve and convolution of our ragged map. Whenever Mom doubted herself, I should have engaged her doubt in conversation. I should have allowed that doubt to become real, and thus something we could change and modify like any real thing. I should have reminded her of her own words. "Assurance is that evasion by means of which cultures exist. The world we seek to grab hold of often grabs hold of us." But I didn't. I believed ultimately that the world was filled with firm and self-evident truths, like those in the Declaration of Independence, and like all people of true vision, both Mom and I would always share these truths, we would always know where they were located. Thus confidently I had allowed my mom to drift away and grow lost in endless dialogues between herself and her own reflection, believing as I did that the world's firmness would always lead her back home to me. So now I deserved it, wasting my days in that insipid school, drifting aimlessly among the rusting climb-schemes of the playground, engaged by my own subjective and watery dejection. I was beginning to feel not only despondent but unreal. The world was growing filled with sharp things, things that banged and brushed against me, things that crowded and pressed me. I, meanwhile, was growing more and more immaterial.

"Phillip," my teacher would ask, "would you like to be the next one to read out loud?"

"I guess so."

"Do you want to learn how to hold a hacksaw?" Pedro's

hand held my shoulder with genuine concern, a concern that threatened to make my shoulder in some way his. "Do you want to learn how to solder metal?"

"I guess so."

"Do you want to help me with dinner?" Mom asked. "Or do you want to go outside and play with your friends?"

They were embroiling me in these unanswered and impossible questions, questions without answers, only compromises.

"Is there anything you'd like at the store?"

"Do you know what you want to be when you grow up?"

"Can you tell us the capital of Delaware? South Dakota? Spain?"

"Who's your favorite movie star? What's your favorite book or television program?"

Blizzards of questions. Questions that infested the air like battering moths, knocking against things, dying alone in blistering glass lampshades among blazing heat and their own aborted larvae.

Perhaps they couldn't control me, but they could limit my ability to control myself. Perhaps my teachers couldn't transform me into some gibbering Audio Visual Monitor, content with my colored paper and chalky paste. Perhaps Pedro couldn't indoctrinate me with metric drills, high-speed power lathes and hammers. Perhaps Mom, seeking to evade her own tragic and naive compromise with the world of Pedro, could envelop me in draperies and new cotton underwear and the radiant warmth of my own portable TV.

22

My secret internal motion, however, couldn't be so easily disavowed. At least that was the mythology I tried to weave around myself like a protective blanket or deliberate dream. They would have to disavow my breath first, my heart, the quality of my voice. Some nights Mom would lie in bed with me to help me sleep, and I would remain stiffly and brazenly awake in her cool arms. I couldn't hear her breathe anymore. While she spoke, I pretended not to listen. While I dreamed, she pretended not to know.

"If it makes you feel any better, baby, I'm not doing all this for you. I wouldn't condescend like that. And it's hard to explain what I'm trying to find here myself, but I do think I've found it. Pedro is a very kind, unimaginative man who never bothers me when I don't want to be bothered. He promises me security, baby, and the deepest sort of privacy too." As she stroked my damp brow I felt the entire universe contract around me. Mom's lies were involved in some vaster scheme of lying. There were vaster deceptions being organized in the universe than Mom's passionless bed with Pedro. "But if you ever want to talk about anything, you know you can tell me, baby. It may not change things, but it might make you feel better. Just talking about things helps sometimes. And then sometimes it doesn't help at all."

But of course I couldn't say anything. That would only betray myself to the mindless airy abstractions of Mom's lustrous deception. I could display only my thin affected drowsiness, pretending as if I too were warm and secure in Pedro's ambivalent home. I guess that's what I hated Mom for most, my own timid and recalcitrant dissimulation. I felt

23

like some burglar or criminal forced to flee the world rather than rush, as Mom and I once had, fiercely into its expanding and elliptical heart. The only freedom they allowed me was to dissemble and resist, to disguise that brisk and fundamental pulse of myself from this false world's pulselessness.

They couldn't get inside me, you see, but they could so alter and confuse my world that I might actually forget how to get back inside myself. It was like Wittgenstein's allegory of the matchboxes. Even though I knew and preserved that special and untransgressed secret of myself from the world's systematic fiddling, ensconced in its immutable privacy the secret itself ceased to breathe and turn. It became an artifact, like something buried in the stale air and glass cases of some shoddy museum, one filled with estranged and obdurate guards in blue suits and official-looking hats that didn't quite fit. I wasn't Mom's baby anymore. I wasn't the rider of Mom's ceaseless motion. I was just another kid in school. I was just a child awaiting his "formative years," coddled with warm blankets and bland, nutritious meals of Wonderbread, peanut butter and grape jelly. I was just a matchbox. I was just a thin matchbox in which some broken object could be heard rattling back and forth. It might be a penny. It might be a plastic green soldier. It might be fragments of a splintery pencil, or a pebble, or a rusty nail, or some scrabbling insect. Or it might be just nothing. It might be nothing worth having at all.

———

My diet, education and serenity were strictly regulated and monitored from now on. I was to go to good movies, read good books, eat nutritious meals, defecate and sleep at pre-scribed hours. I received a haircut at the barber's every two weeks. I received inoculations for polio, tetanus, smallpox, diphtheria. I suffered a visit to the dentist, where a cruel hygienist scraped the hard crusty plaque from my teeth with sharp steel instruments. "You're very lucky not to have any cavities," she told me, and I could only think, There, I told you. I never needed you to begin with, as I spat blood into the white bowl's blue, cascading water. I received stacks and stacks of new clothes, though my drawers were already filled with freshly pressed and laundered shirts and slacks. My old friendly Levi's and sweatshirts vanished while my closet blossomed with toys in boxes, colorful books and sports equipment, flashing electronic games and educational video cassettes. "You know, I was thinking," Pedro said one day, refolding his paper and placing it in his lap, pulling off his sparkling bifocals with a flourish. He gazed blankly at the ornamental knickknack shelf he had installed just that eve-ning. "You know what Phillip needs? Phillip needs a dog. A nice little puppy he can raise and take care of. It will teach him a little responsibility. It will be his good friend whenever he feels dejected and alone. If he keeps it well brushed and groomed, he can even let it sleep at the foot of his bed. I don't know why we didn't think of that ear-lier," Pedro said cryptically. "A dog."

And then, wordlessly suffering on the carpet with my

schoolbooks, raptly gazing at the heatlessly flickering television, I could only listen as Mom concurred with an earnestness which made me sick to my stomach. I felt deep intestinal kicks and grinding. Heat lifted in my heart, my chest. The blood rushed to my head and I felt dizzy and slightly nauseous, as if I were ascending into the high air on some sudden spaceship. We will go to the pound on Tuesday, Mom said. No, Wednesday, because Tuesday I work. We'll get a license. And Pedro, honey, you can build a little doghouse in the yard. We'll make a little mattress inside with old rags and things. When it's potty trained, we'll even talk about letting it sleep at the foot of Phillip's bed. We can buy books about training dogs, dog grooming and health care, dog dogness and doggish dogs. Bland little puppies which you hold in your arms like presents. They all have big floppy ears, big soulful eyes. They always love you, no matter what. No matter how you feel about yourself, dogs think you're the greatest. No matter how harsh and insincere the world is, dogs aren't. Dogs love you even when you kick them, even when you don't feed them. Dogs love you even when your hands clench their throats. Dogs love love love you even when they can't breathe, even when their tiny soulful eyes grow more bloodshot and confused with actual terror, even when they give that final, galvanic little kick and their breath stops. When they grow rigid. When their eyes turn glassy and reflective. When you bury them in the garden with a tiny wooden cross and pray for God to forgive them all their sins.

4

With all my polluted and forlorn heart I prayed Mom would shamelessly murder me just like that, the same way I would surely murder any conceivable puppy with which they might seek to burden and restrain me. Kill me and let it be over with, I prayed each night in my feverish bed. Kill me with your own hands so I know it's you. Like the puppy, I will still love you; I will never stop loving you. Like the puppy I will trust you always and forever, right up to the very end.

I felt weaker every day, more listless, distracted and pale. Mom, though, never seemed to notice. Vulnerable and more diffident, smaller and smaller, I was drifting further and further away from her like some eccentric planet. "Culture has a valid purpose," Mom said, seated on the verge of my bed with her cool hand in my lap, abstractly gazing out the window at the red apocalyptic sunset, distantly contemplating the intricacies of her own subversion. "It's not like culture's out to get us. It's not like we have anything to fear from culture but ourselves." Sometimes, as I stared at her,

27

her voice seemed to grow dimmer and more diffuse. I was beginning to realize that Mom had not succumbed to the world's lies, but rather to the sudden swerve and convolution of her own extraordinary mind. "Culture's just a scheme of rules and regulations we've all quite happily agreed to. It's not like all the clichés, baby. Like we submit. Like we're oppressed or imprisoned or enchained. Culture's got our best interests at heart. Culture's just the walls of a house. It's that house I always told you we lived in, only I didn't realize that house was culture before." Mom was wearing a slightly ragged and pulpy white nightgown which had belonged to Pedro's deceased wife, Marjoree. Mom was gaining a little weight; her beautiful face had grown slightly pale and flaccid. The palms of her hands felt cold and dry. "Freedom is a place inside your own mind," she said. And now we were in different galaxies, Mom and I, spinning among remote civilizations and suns. "Culture's just a set of rules that makes life comfortable. That give us time for the freedom we can only live inside ourselves."

Mom said I was suffering growing pains. Pedro was the one who began bringing home the doctors. My temperature was taken, my blood and pulse. My malaise was misdiagnosed as influenza, trauma, shock, diabetes and once even leukemia. I never bothered to get out of bed anymore. Letters were written to and from my school; a tutor occasionally arrived and sat beside my bed, as cold and indifferent as Mom with his mundane assignments and texts. I spent all day watching TV. The morning news. A few hours of game shows in which the world's insipid and luxury-starved eagerly competed for new washing machines, trash compactors and automobiles. Perhaps a soap opera or two and then, finally, the talk shows. Mike Douglas was my favorite, but there were days when Merv Griffin was my favorite too. I liked the talk shows because they featured a revolving panel of guests who had just flown in from limited engagements in Tahoe, Reno and New York. They all had many stories to tell, most of them amusing and comfortably inconsequential. They knew that language was a sort of padding or excess. It was uttered with practiced enthusiasm. You could talk and talk and talk on TV and never have to say anything. I lay in my bed and never said anything either.

Where Mom had once lived her life in the world she now lived her world in the mind. It was a secret world filled with dark speculations and sober intricacy. Vast and comprehensive theories were worked out down there, enthralled by senseless reason. Complicated chiaroscuros of reflection like magnificent Venetian tapestries. Extensive logarithms of interpretation like sculpted white clouds. Mom's secret self sat there in its immaculate kingdom, merely dreaming of other kingdoms like mine. "We've started your college fund," Mom said, enthroned on the edge of my bed, her cool hand petting my sweaty brow. "Pretty soon we'll start looking into a few of the better prep schools. When Pedro retires in a few years we'll look for a larger house. You can travel in the summer. You will always have a home to come to, always a little money in the bank. Then you'll be free to be anybody you want to be. You can go to medical school. You can be a rock star. You can be a stage actor or a vice-president. You can shoot drugs or hire hookers. You can become homosexual or a hired assassin. It's your life, baby, and you live it anyway you choose. I'll always love you, no matter what. Just always remember—you need to play the game if you want to break the rules, and even if you play by all the rules, deep in your brain you'll always be playing your own game. You are immaculate. You endure for numberless centuries. You persevere in a world of pure gravity and sound. You are like light, baby. You are like a sea of air. You are history, and make all of history something else."

I could hardly sleep at all anymore, tossing and twisting among my feverish sheets, hearing Mom's steady breath in the bedroom adjoining mine, Pedro's own antiphonal and staggered snores. When I did sleep I dreamed I was awake. I dreamed Mom was sitting on my bed. I dreamed Pedro was building and hammering in the backyard. I dreamed the teachers and other schoolchildren were telling me I had done a very good job. I was easy to get along with, they all liked me better now that I tried, now that I made some effort to be fun to be with. A spectral puppy licked my face. The ghosts of my delirious life assembled around me even in the dark, compensatory and half-lit world of my dreams.

Pedro built me a sturdy lap tray so I could eat my meals in bed. He built me a bookshelf and a large wooden toy chest, and subsequently filled them with board games, jig-saw puzzles, woodburning and constructor and Lego sets, sacks of green faceless plastic army men, a variety of baseball mitts and a solid, unscratched hardball, seamed and dense. My room was filling up with more and more weight and mass. I could feel the foundations of our ranch-style open-plan house beginning to creak and kneel—I and my room full of things poised to abrupt through the floor, through the earth's crust and mantle, rejoining that infinite and unseen history of strange misshapen creatures with rattling carapaces and stunned, minuscule brains. I was nothing but pure weight now, hard matter. I couldn't move; some nights I couldn't even breathe. "You stay in bed all your life if you want to," Mom said, after the bespectacled psychiatrist suggested I go away for a few months. He was a member of the advisory board to a "special" ranch where children like me conventionally responded well to treatment. This hypothetical summer camp was filled with ponies and swimming pools and campfires; young boys and girls of my own age slept in tents there, sang campfire songs and traveled down rivers on rafts. But Mom wouldn't let them take me; Mom told them I would be all right. That was my mom. Even while she was destroying me, she would take care nobody else destroyed me. "There's nothing wrong with a few months of uninterrupted reflection," she told me that night. "As long as we're happy. As long as we're all happy in our house together, there's no reason why we should be in any rush to go anywhere."

5

I knew where they kept the Seconal. On the top shelf of
the master bathroom's medicine cabinet in an amber child-
proof bottle. You had to depress and crank the lid of the
bottle with the heel of your hand. Then you held it there,
a sweet unconscious turning in oval gelatin capsules. Sometimes I might take one, letting the capsule dissolve on my
tongue, tasting the grainy barbiturate seeping through, bitter
and full of life. Then I would replace the pills in their container and step quietly back through Mom and Pedro's
room. Mom lay on her side, facing me, her eyes glassy and
volitionless, watching me without deliberation, permitting
me at least my secret life. I saw myself reflected in her eyes,
and the moonlight where we converged. "I'm going to do
it, Mom," I whispered. "I don't want to make you unhappy, but I think about it every night."

Mom didn't say anything. Perhaps, absently, her right
hand might gently stroke her left shoulder. Her dark eyes
might turn and follow me to the door.

"I know you'll let me," I told her. "I just want you to know. I don't want to hurt you, but I can't stand to let myself be hurt any more, either. If it was between you and me, Mom, you'd make the same decision. You'd always choose to hurt me rather than hurt yourself." And then I crept silently back to my room, tracking a spoor of glowing red ash across the carpet of Pedro's dreadful house, dreaming my inviolate dreams of motion again. In my dreams I was moving, with or without Mom, across lawns and galaxies, streets and stars, suburbs and unraveling solar winds. The Seconal was my ticket out, and I was going to use it.

Mom was working late that night due to a last-minute change in her schedule.

"Hey, sport." Pedro was watching a Dodgers and Giants game on TV and drinking his customary Budweiser. "Out of bed tonight, I see. Good game here, if you want to watch it."

Mike Marshall had just fouled a hard sinking fastball off his right foot.

"Now Marshall's walking away from the plate and *boy* does that smart," Vin Scully, the announcer, said.

"Used to play a little pro ball myself." Pedro was digging into his ear with the little finger of his right hand. After he was finished he shook his head slightly, as if he heard something rattle inside. "Some double-A ball in the Texas League. That was back in sixty-two."

It was funny, because suddenly I didn't even hate Pedro anymore. In fact, as I sat and talked with him that night,

33

the world of menace I once associated with Pedro's name seemed to withdraw a little. Grow lighter and more gaseous, its molecules quicker and more excited. Pedro. I was suddenly convinced of the fact that Pedro *was* a very nice man, and that conviction filled me with an impossible sadness.

"That's where I got my nickname, you see." Pedro's glazed eyes dimly apprehended Marshall on first, Guerrero on second. Atlee Hammaker was pitching for the Giants. I really liked that name. Atlee Hammaker. "I never said a lot when I was a kid, and everybody thought, because I had really black hair back then, that I was Mexican. I really did look Mexican. I looked about as Mexican as you could expect a Mexican to look." Pedro ran one hand through his gray and thinning hair. Hammaker struck out Bill Russell on four pitches. "Damn," Pedro muttered. "Damn it, Billyboy."

"Sometimes you just have to make the effort," Pedro consoled me later. The game had gone into extra innings, tied 2-2, and Pedro had turned the volume down by means of his remote control. He was on his sixth or seventh Budweiser, and I was preparing to fetch him another from the fridge. "I mean, it's not like I ever had these big *ambitions*, you know, to run a *hardware* store, for chrissake. I mean, opening a hardware store wasn't something that, you know, woke me up excited every morning. Like I'd wake up thinking, *Hey*, I own a *hardware* store! *Hey*, I'm on my way to work in my very own *hardware* store! Hell, no. It wasn't like that at all, kiddo. I mean, running a hardware store was just a lot of hard work every day, believe you me. There were plenty of days when I just wanted to lie in bed too. No lie. I would have loved to just lie in bed and watch TV and listen to ball games on the radio. But back then, you see, I couldn't afford to hire any help, and if *I* had stayed in bed all day, just who do you think would have run that hardware store? Who do you think would have paid my mortgage so I *could* lie in bed all day? Nelson Rockefeller? Think again, kiddo. Howard Hughes—my good old buddy Howard? Well, I doubt it. I can't say for sure, but somehow I doubt my old buddy Howard *Hughes* would've come round to help pay off *my* mortgage."

He opened another beer, and I warmed some canned chili on the stove. Pedro ate most of it, sponging up the red chili sauce with slices of his doughy Wonderbread. "This is a hard fast world we live in, kiddo—and I'm telling you this as a friend, now. All this teary-eyed feeling sorry for yourself *childhood* crap just doesn't work—doesn't work for long, anyway. I can promise you that. I mean, your mom wants you to have this *idyllic* childhood and all. She thinks this is Camelot or something, your childhood. Well, I want you to know, kiddo. I looked up 'idyllic' in the dictionary and I wouldn't hold my breath. I wouldn't lie in bed all day just waiting for some idyllic childhood to come along."

I know, I wanted to tell him. You're right. Love often requires sacrifices which simply aren't worth it.

"So maybe you've had a few hard knocks. So maybe you've lived a sort of fly-by-night existence and all. That's just the breaks, kiddo," Pedro said, and for once I listened. For once I wanted us to hear each other. "That's just life.

35

And believe you me, we sure live it a damn sight better than we do lying in bed all day feeling sorry for ourselves. I think that's the truth, kiddo, and . . ." He gave a tremendous yawn. "Jesus." He blinked his eyes. His crumpled Budweiser cans lay toppled around him on the table, sofa and floor like crude chess pieces. "Boy. I guess I'm really bushed." Pedro pushed himself to his feet, slouched, pot-bellied and creased by the rough sofa cushions. "You take care of yourself, kiddo," Pedro told me, and shuffled in his wrinkled suede slippers toward the master bedroom. "I think I'm going to hit the hay." And then I heard him groaning into the squeaky bed, drifting into his slow aimless dreams of the soft red barbiturate haze that filled him like warm air fills a balloon. Meanwhile the ruptured gelatin Seconal capsules lay scattered on a sheet of Kleenex on the desk in my room.

I finished my diet soda and went in to see how he was doing. He looked very warm and peaceful, his face flushed and puffy, his vital bodily signs sailing along gently and intrepidly and slow. All the long steel kitchen knives were unsharpened and dully glimmering in the kitchen cabinets. There was no heavy cord or rope anywhere to be found, and though I suspected there might be some in the basement, it was dark down there, cold and damp, and I wasn't wearing any shoes. Then, like weather, I felt it first, just the heavy simplicity of it, a faint steel resonance underneath Pedro's bed. For a while I stood there and appreciated that strange, almost tactile presence. It was very solid. It was very useful and perfectly designed. Clearly it would do the job.

After a while I pulled Pedro's toolbox out from under the bed where it waited for me like history. I lifted its impossibly heavy weight onto the foot of the mattress. The toolbox contained hammers, screwdrivers, ratchets, Allen wrenches, hacksaws and spare, gleaming new replacement hacksaw blades. I knew that Pedro wanted a world as secure as the things he constructed in the backyard, a world with perfectly articulated joints and level, sanded surfaces. I knew that Pedro deserved a world like the worlds he built out there, like the worlds he built inside himself and Mom. "Death is the hard song, Pedro," I told him. "We only sing it once, and none of us ever gets it exactly right."

Even as I inaugurated my secret ceremonies of redemption that night, I knew something vaster and more important than myself was responsible for all my actions. Me, Mom, Pedro, and Mom's vast world were all just fragments of a process that would soon consume us all. I didn't want to give in to that mindless process, you see. I wanted to leave something behind, like the pyramids in Egypt, or the heads on Mount Rushmore. I wanted to build something formidable and good for all of us, but especially for Pedro. All that long night as I feverishly worked, what I wanted to do more than anything was build something for Pedro that would last forever.

LIGHT

———————————————————

6

I thought when Mom saw what I had done to Pedro she might stop loving me, but from that night forward I think she may have started loving me even more. When she emerged expressionless from the master bedroom I was sitting on the living room sofa, gently stroking my wet clean hair with a brown towel, still stippled and muggy after my long mournful bath. She didn't pause or speak to me. She just began packing our few belongings into pillowcases, and after a while I dressed and helped her carry everything out to the garage where our old Rambler had sat gathering dust and ticking these many months, sluggish and thick with its own unstirred oils and rusty water. It started up on Mom's first try. Then she held down the accelerator for a while and we sat there sleepily in the dark garage, staring out at the brighter and more opaque darkness beyond the roar of our Rambler's V-8. We were lifting off. In a moment, we would be hurtling through space. Mom released the emergency brake and the V-8 subsided to a rough, hesitant idle.

Then we glided down the long cement driveway while Pedro lay asleep in his calm and remorseless home, dreaming his dreams of barbiturates, beer and the soft biting blades of tools and things. God, I was filled with light that night. I was filled with Mom's voice and the very light of her. We were moving again. We would never die. We would travel together forever in the world of inexplicit light, Mom and I.

"The history of motion is that luminous progress men and women make in the world alone," Mom said. "We're moving into sudden history now, baby. That life men lead and women disavow, that sure and certain sense that nothing is wrong, that life does not beat or pause, that the universe expands relentlessly. You can feel the source of all the world's light in your beating heart, in the map of your blood, in the vast range and pace of your brain. That's the light, baby. You don't need any other. Just that light beating forever inside of you." We were turning onto the freeway, which was filled with other, hurtling headlights, enormous menacing trucks and buses. "We are like astronauts, we are like wheeling planes and spaceships. We are like swaying birds with soft stroking wings like oars. We beat against the heavy air, and carry our silent and regenerate light with us wherever we go."

It was nice Mom telling me that, that the light was mine too. But I knew the light was Mom's and nobody else's. For months I had seen nothing but my own interior and subjective darkness, and now, against the glare of Mom's resumed motion, I could see the entire world again. No, all the light we gathered was Mom's light, Mom's progress into places I could only dream about. I was just a passenger, and like all passengers, fundamentally unconcerned with landscape and plot, enveloped only by the simple movement of it all, the cumulate graph of those coherent points where we ate, slept, went to the bathroom, and awaited movement again. We could live together forever and ever, again and again, life after life. Mom didn't have to lie anymore. She didn't have to run or hide, she didn't have to journey further away from me in order to remain with me as she did, deeper into her dreams of me and further away from my untrained arms. I didn't know it then, but I was soon to learn I couldn't follow Mom everywhere.

These days I was intent on immortality, because I knew Mom's only hope of redemption lay in time itself, some expansion and unfocusing of time that would swallow Mom and all her imaginings into one formless shape and sound, not a place so much as a force or dispersion of force, an abstract location that bound and contextualized things, like gravity or sound. "Low-cholesterol diets, Mom," I told her, browsing through a college nursing text entitled *Health and Our World: 32nd Edition.* "Then there's the DNA, those complex looping signals beeping in our blood and our lymph. Death's a program, Mom. Like eating, sleeping, sex and hate. Our bodies generate death like fluids, waste, carbon dioxide, anticoagulants, marrow. DNA's the beeping clock, unraveling time in our bodies like smoke from your cigarettes. It's the tiniest force; it responds with information, not blood; it circulates raw and genetically contrived data, not life exactly. The heart—we'll leave that to the regular scientists. There's some oils in fish that cleanse the body of fatty tissue and keep the rich blood pumping. But down into the DNA is where I'll go, Mom. When I grow up I'll have a laboratory. I'll invent a lot of stupid consumer junk so I make lots of money. Then I'll sink everything I've got into the DNA. I'll climb down into its bristling helical nets like a spelunker. I'll dig out every secret, and then they'll be our secrets, Mom, and we'll live forever. We'll have a house overlooking the beach, and my laboratory in the basement. And we'll live together without anyone bothering us for thousands and thousands of years."

Most of the time Mom just drove without looking at me, wearing her tortoiseshell sunglasses and a floppy straw hat. She was listening, somewhere deep in her brain, but she was watching other roads now besides 101, other routes besides the one on a map. "This is King City," she might say. "I think we've been to King City." Mom's face was very pale without makeup, but very beautiful as well. "Let's try it anyway," and pulled onto the next off-ramp. Soon we were winding down into a Burger King, a Wendy's, a Motel 6, a King's Bowl Bar and Grill. I always insisted on a salad bar in these days of Mom's growing disaffection. I urged her to eat plenty of raw vegetables and fresh fish. We would pull into the parking lot and she would turn to me. "It's got to be better than San Luis, doesn't it? It's got to be better than that hellhole." Then she gave the fleshy thigh of my arm a little squeeze and smiled. Only she wasn't looking at me in a way. She was looking at me, but she wasn't looking at me at the same time.

Rather than disappearing into neon bars with her strange, generally unmanicured men, Mom took longer and longer looks at herself in the vanity mirrors of our motel rooms, drinking her Seagram's and 7UP, her Scotch and Tab. She would wear her laciest lingerie and just sit there alone. Perhaps she would paint her face with very bright makeup, or contrast her pale cheeks with soft blushes and eye shadows, leaning forward, one elbow against one dimpled knee, one brilliantly manicured hand splayed gently against the top of the dresser, her other hand producing various vials and Maybelline from her handbag, which bristled with crumpled Kleenex, tattered road maps, plastic cutlery, and the various salt, ketchup, and NutraSweet packets she had lifted from fast-food restaurants. Her breasts were fully outlined against the sheer fabric of her lingerie; her long, slightly pudgy thighs (of which she was curiously ashamed, and over which she generally wore pants or thick cotton "middie" skirts); her legs glistening with dark nylons. Sometimes, as she watched herself applying makeup, she might take a few long slow breaths. I could feel her breath warm in the air around me; I could taste its warmth against my skin and face. Sometimes her nipples grew more prominent and stiff. She would remove her left hand from the table and place it against the inside of her left thigh. Lying on my side of the bed I watched her, and my body filled with strange, smoky sensations. She wasn't looking at me. She wasn't looking at me. But I was looking at her.

I began to feel a little out of breath, resting the open textbook against my thin, almost concave chest. Mom was a bird, a cloud, a car. Mom was something that breathed like me, that felt warm like me, that could move her legs like mine. She wasn't looking at me, but I was looking at her. Her face emblazoned with cosmetics, her body firm and distant and unbelievably warm. I was becoming her only man. No other men ever came around. I was watching Mom and, after a while, out of the corner of her eye, Mom began watching me, her hand which held the lip gloss hovering against the edge of the dresser, her cool gaze turned in my direction now, as if she saw me and she didn't see me, and I felt my entire body burning and pulsing with the light, the light, all the night's darkness which was now suddenly turning into light, and all the sleepiness pulling at my face and filling my eyes with heat and softness and a sort of blurred and amorous detachment, and then I was falling asleep, and my body gave a sudden little kick. And as I slept I dreamed of Pedro. I dreamed of Pedro dreaming of me. Because Pedro and I understood one another perfectly now. We both loved Mom. And now we were all that was left of that strange and delusory world of Mom's numinous men.

Many of our surviving Visa and MasterCard cards were beginning to reach and overreach their expiration dates, and Mom and I soon grew stingier with our fund of invisible credit. We began pulling "runners" at restaurants, coffee shops and motels. While Mom flirted in the office with mechanics and gasoline attendants, I would jimmy open the cash box out on the service island with a screwdriver and pull out the large bills from underneath the steel change tray. We lifted food from grocery stores and clothes from clothing stores. We took magazines, beer and cigarettes from 7-Elevens, Stop 'N' Shops, Liquor Barns and Walgreen's drugstores. One afternoon at the Van Nuys Motel 6 I was returning to our room after playing one of my slow games with a sharp stick and a dead, forlorn blackbird, and found Mom carrying the color portable television from our motel room downstairs to our car. We sold it that night to a pair of diminutive and portly Mexicans—very pleasant and smiling men, as I recall—for twenty-five dollars in the parking lot of Serra Bowl in Encino. "Value's generated by the world, not consciousness," Mom said that night as we drove south to La Jolla. "The trick is to take the world and its values and generate better worlds inside. You've got a choice, baby, and it's the only choice you've got. Either remake the world, or allow the world to remake you. Did that sign say 101? Look for my glasses—there, on the dash. And keep an eye out for Highway 101."

We were driving, always driving, and always it was night. Outside our hurtling car the darkness simmered with radio waves and the swirling, hot Santa Anas. Everything converged out there, even the heartbeats of other stars and galaxies. Pulsars, quasars, fissioning novas and supernovas, the radar of airplanes and control towers, the diminishing cries of hidden and crepuscular birds. I couldn't look out into that eternal night—those inconstant oceans of static engulfing our AM radio every few miles or so—without thinking the question. The question surfaced like some underwater creature. It was learning to oxygenate. It was crawling from the sea's boiling muck.

"Whatever happened to Dad?" I asked Mom, against my will. I couldn't help myself. The question was like force, blood pressure, chemistry or light. "Where is Dad now? Is he still alive? At night, like this, when the night is just like this, does Dad ever think about us? Is Dad a person in the world, Mom? Or does he just lie in his bed and dream? And if so, Mom, are we his dream, or is he ours?"

But Mom had already grown very quiet. It was almost as if the question were not mine at all, but rather part of some thin formless lapse within the continuity of Mom's diminishing world. She never said anything for hours at a time. I began to realize my mom was going very far away. I merely traveled, but Mom journeyed.

7

Then one day I awoke puffy and unbathed in the backseat of our car and Mom told me. The hot sunlight was filling the cracked vinyl upholstery, the warped, discolored dashboard and dirty windows. Mom was leaning inside and pushing my shoulder. "I've done it," Mom said. "I've rented us our own house." So finally, after years and years without memory, Mom inaugurated time for us again. We had our own house now, and nobody lived in it but us.

"I think I can say I've learned a lot of important things in the past few months or so," Mom told me that night. "About myself, and you, and our world, and our future. And about the sort of unrealistic expectations people can have about one another. Everything's going to be a lot different, this time," Mom promised. "I think I've learned to be a little more realistic about things. I've learned there are some things we simply can't expect from one another."

Every few minutes she took her glass into the kitchen and hacked at the bag of ice we had purchased that evening from the neighborhood liquor store. The ice rested in the rusty and chipped Formica sink, thawing and reshaping itself. Then Mom returned to the living room with her ice-filled glass and poured more Seagram's and 7UP.

"I don't care, Mom," I said, compelled by my own confessions too. "I just want you to know that I'm not mad at Pedro anymore. I think I may have been very selfish and confused lately, and I don't mind if you want Pedro to come live with us again. I know I can't keep you all to myself, because my love for you can't be a selfish love if it's to be honest and true. I know now I have to let you live your own world, because that's what I love about you. That world you are apart from me. I think I'm beginning to learn a lot about myself as an individual, Mom. And if Pedro comes to live with us again, I promise I'll be nice to him this time. I promise I won't do anything I shouldn't do."

Mom sipped her drink in the cold room, the candles flickering around us, impaling the mouths of Mountain Dew and Coke bottles streaked with ruddy wax. Mom just looked away. It was as if she didn't hear me. It was as if she were listening to Pedro dream, the man whose name she taught me once to say and then taught herself never to say again. I wondered if in Pedro's dreams there were visions of Pedro dreaming, like the way angled mirrors reflect one another infinitely in department-store dressing rooms. In Pedro's dreams there was Mom, me, and a dark gathering shape underneath the floors of our new house. The dark shape said, "The family environment is a very important place for growing children. A stable family unit environment determines whether a young child will grow up feeling assured and self-confident, or simply undisciplined, slothful and insecure." Whenever we heard that voice coming, Pedro always shot me a glance of warning. Pedro and I both knew Dad would be with us again very soon.

"Sometimes it's hard to tell the difference between your conception of the world and the world's conception of you," Mom said, swirling thin ice in her glass. We slept on the shag carpet on the rolled-up blankets and quilts we had lifted the previous evening from the Best Western Motel in Van Nuys. "It's very easy to fool yourself," Mom said. "The harder you think about things, the more confused you get." She was lying on her back and gazing at our white, water-stained ceiling. Her hands were resting very quietly on her breathing stomach. "When I was just a little girl I would sit on the living room couch for hours sometimes, just trying to figure out the simplest things. I couldn't even move. My mind grew fuzzy and dim. I felt as if my skull was inflating with chemical pressure or anesthetic. It grew dark outside. My mother returned home from work. She fixed me dinner, but I wouldn't eat. I just sat there alone until I could feel this sort of moving black cloud slowly engulf me. Inside the black cloud, I couldn't think about anything. I couldn't even remember what I had been trying to figure out. Sometimes I couldn't remember my own name, or the address where I lived. Sometimes I couldn't even be sure if my mother was really my mother at all." Stealthily, the gas heater gave a tiny kick in the kitchen. Outside, the city was filled with bright, airy noise, whisper-

ing against the thin walls of our house like something cor-
poreal, filled with hissing and irreducible life.

"Go to sleep, Mom," I told her, and placed my hand on
both of hers. "Get some rest and we'll talk about it in the
morning."

"Sometimes, lately, I've started feeling like that again,"
Mom said. "I see this cloud of blackness moving in around
me. I start to forget things. I can't even tell if I'm dreaming
or not."

Outside in the bright night, the full moon gazed over
everything, gravid with implications which, at my still tender
age, I could suspect but not yet comprehend.

"Your father took me away from all that," Mom said
distantly, "and that's why I'll always be very grateful to him.
I'll always be very grateful to your father, Phillip, but that
doesn't mean I want him back."

Mom always said we would buy furniture someday, but we
never did. Instead we purchased a Hitachi color television,
VHS recorder and remote control with one of our few
remaining credit cards on which time, like the vital current
of some living creature, was gradually running out. We purchased a pair of springless Sta-Easy mattresses from a ridic-
ulously exorbitant Salvation Army thrift store and placed
one in each of our musty, isolate bedrooms. We purchased
an audiocassette recorder and various new tapes from Tower
Records in Van Nuys, and a small unvarnished desk with a
built-in bookshelf for my room, one on which I assembled
my various stained and pulpy textbooks, a new notepad,

pink rubber eraser, plastic ruler, pencil sharpener and pencil case. These were my tools now, and like Pedro I kept them all in their proper place. There was something submarine about them, even anxious. Mom had recently determined I would be a writer someday.

"Take words and make them useful," she told me. "Drain them of all the crappy meanings they *used* to mean, and make them mean something useful instead." I assigned myself to my room for exactly two hours every morning, where I studied my books and wrote my clean words. With my elbows propped against my splintery desk, I plunged into books and histories and explicable mysteries like some hungry and lucratively sponsored wilderness explorer. I made vast new areas of knowledge cultivable and known. I descended to the ocean floor and encountered bloated, symmetrical creatures with pumping white hearts and translucent skin. Collapsed blue civilizations lived down there, fissured and antiseptic, craggy with barnacles and blistering rust. I reached into the heart of the earth, the sky, the moon. I colonized language, mathematics, schemes of chemical order and atomic weight. I studied the manufacture of automobiles, microcircuitry, Kleenex and planets. I memorized the gross national products of nations and hemispheres, the populations of cities and states and principalities, the achievements of presidents, tyrants and kings. I was trying to learn what I suspected Mom had learned already: that there were journeys we all make alone that take us far away from one another.

Every morning I awoke alone in our cold house and padded softly into the kitchen, where I prepared myself Pop Tarts, hot chocolate and perhaps a small bowl of cold cereal. Then I would turn on all the stove's gas jets to break the kitchen's chill, and sit at the wooden breakfast nook perusing last evening's *Herald* (I disdained the *Times* for political reasons). I might listen to a little local all-chat radio for a while, and then fix myself a small pot of coffee and return to my study, always attentive as I passed Mom's silent room, where she remained discreetly asleep or self-absorbed until around midafternoon. Then I read alone in my room until at least noon, spilling the strange energetic words into my head. Geology, psychology, ancient history, applied linguistics, German, modern philosophy, South American etymology, Central American politics, Fourier, Rousseau, Marx—a vast boil and suck of words and languages. I recall very little of what I learned then; the ideas didn't really cling. Rather they seemed to seep into my skin and belly and condition rather than fill me. It was as if I were just modifying the shape of my hunger rather than appeasing it. The only knowledge that really mattered to me then mattered because it was linked somewhere in my feverish imagination with the emerging shape of Dad. I remember quantum physics because I felt that Dad, like the movement of planets, was not a fact of data so much as a quality of interpretation. I remember European revolutionary governments of the eighteenth century because their subversion of "Father" had never eliminated so much as merely redesigned his very real presence. I remember Hegel because I always imagined that the thisness which was Mom and I was always transforming itself into the thatness which would be life with Dad. Dad was the thatness towards which all our complicit motion yearned. It was in February that he called for the first time.

Is this Phillip? he asked when I answered. I had never answered a phone before to the sound of my own name.

"Who's this?" I asked, but I didn't really need to ask. There was only one other person in the world who knew my name.

This is your dad, he said. This is your dad who misses you both very, very much.

I hung up, and he didn't ring back. At least not that same night.

Later when I went to bed I tried to distinguish the different schemes of light that infiltrated my room. There was the lunar and the electrical, the stellar and the reflected. There was the light of ghosts, and the light of living things. That night Pedro spoke to me for the first time since he began dreaming of those hard lightless objects which filled his somber toolbox.

"I forgive you what you did to me, but I'll never forgive you what you did to your mom. I'll never forgive you what you did to yourself."

"But what about the light, Pedro?" I asked him. "What sort of light do you see now? Does the light that fills you make you feel warm, or safe, or sad?" But Pedro's voice had grown silent again. He had said all he wanted to say. It was as if, while he dreamed, someone was keeping watch over him. Dreaming was a prison of some kind in which you were never really alone even for a minute, in which you were responsible to a legion of regulations, timetables and personnel. I couldn't understand why Pedro said he could never forgive me. It had to be a code or a cipher of some kind. If he told me the truth about his new life, he might get himself in serious trouble with the people who gauged and monitored that life. I would have to ask him about it later. For now, effortlessly, I could only sleep.

8

Dad called the next morning around ten thirty.

It's been five years, he said. You guys are a hard act to follow, but not so hard to trace. I may not always know where to find you, but I always know where you've been. In fact you've left a trail which you might say is at least a mile wide. Hearing my detective agency's progress reports on your travels is more fun than watching television, and there are some pretty good programs on television these days, or so I've heard. I was worried when I learned you'd stopped again. I was worried when I learned you'd had the phone connected in your mom's name. The electricity and the gas.

Dad wasn't a voice at all, not even on the phone. Dad's voice surfaced in my life that day like something vaster and more comprehensible than speech, like language itself. I tried to convince him he must have a wrong number, but Dad wasn't buying.

Your mom's had a lot of tough breaks in her life, Dad

continued. (I couldn't imagine what Dad looked like, but I could distantly envision his large body out in some nondescript backyard wielding a long green garden hose. He sprayed the grass and flowers, the easily contented trees and saplings. Then he filled a large plastic bucket with soapy water and went out front to wash the car.) Your mom is a very good woman who doesn't always do very good things. She's not really what I'd call an appropriate role model for a young boy. I think what I'm trying to say, Phillip, is that it may be time for you to come back home and live with your dad again. We'll fix up your old room. We'll get you enrolled back at school. Your mom's welcome to come back home as well, Phillip. I still love your mom, no matter what she's done. And so far as I know, she's done some pretty bad things. There was that poor fellow, Bernie Somebody-orother, in San Luis Obispo. And then, a year or so earlier, that architect in Simi Valley.

I felt a cold breeze moving into my legs, my buttocks, my stomach. It reached into my chest.

"What architect?" I asked. Other worlds were opening themselves to my inspection when I was seven years old—not just the worlds in books, not just the worlds in words. "What architect in Simi Valley? Did he have a red beard?" I asked, not remembering so much as describing him to myself, as if I were the one making the world real with my voice. "Did he have a deep basso profundo singing voice? Did he drive a brand-new green BMW?"

———

Light

Dad called every afternoon and told me things Mom had done. Felonies, assaults, mild flurries of misdemeanors and unacknowledged traffic citations, suspected manslaughters in Burlingame, San Jose, even Whittier. Mom was becoming even more glorious, transubstantial and unreal. She was moving further away from me and into the realm of raw, undifferentiated nature. Mom was a bat, a wolf, a bear, a tiger. Sometimes, as I grew to love her even more, I began to imagine she was luring me into the nests and secret networks of her own convoluted self. Alone in my bed at night, I heard myself talking like her, my mind working like hers. "The irregularities of the world's body correspond with the map of our own brains, baby," I said in my dark room, entangled by my dark and muddled blankets. Gently my hands stroked my stomach, my thighs, the stray black hairs beginning to emerge on my breathing chest. "We travel across the world and into the ways representation works. Trees aren't trees, roads aren't roads, moms aren't even moms. The history of motion is that luminous progress men and women make in the world alone." Sometimes I couldn't even remember which words were mine and which words Mom's. Whose voice was it, whose tongue and whose lips? Where did my flesh of words end and Mom's words of flesh begin? Was this Mom's face and stomach and beating heart, or was this mine? Was I becoming her, some mere reproduction of Mom, or had she so totally and unselfishly invested herself inside me that she no longer really existed at all? I tried to tell myself that I was still me and that Mom was still my mom, but never with any conviction. I am myself, I whispered again and again in the dark. I live my own life. I imagine my own worlds. That's what I kept telling myself.

"Mom's been arrested for soliciting," I told Pedro one night. "That means Mom slept with men and they paid her. She didn't just take money from men, she engaged in business relations. That means they took something from her too." Pedro was dreamily envisioning a new redwood knick-knack shelf with Y joints and notched shelves. He was twiddling his thumbs in his lap just like a little boy. "Mom has been committing crimes I didn't even know about. She has stolen real cash and valuable cars. She even sold hard drugs once. She put two men in a hospital, and at least one in a morgue. Mom has been committing these secret acts without my help, because she's got a terrible temper and she can't help herself. What's more, Pedro, Mom can be cured. Her condition is something that can be altered by the proper medication, regulated by trained doctors and commercial, cost-effective therapy. Mom has a very bad temper, Dad says. Mom has a very bad temper, and I've never even seen it." I was feeling very hot and flushed. Something gave in my stomach, like a loose floorboard. I started to cry. "Mom's someone I don't even know at all, Pedro. That's why I'm growing up so wild. That's why I'm doing things I really shouldn't do. Perhaps that's even why I did the things I did to you, Pedro. But I can't remember. I can't even remember what I did to you anymore." I tried to stop, but I couldn't stop crying. The atmosphere of my small room turned moist and clinging. I felt as if I were crying in the womb of some hibernal animal.

"If you're gonna play hardball, you're gonna get hurt," Pedro said wisely, drifting away into the mist. "We're all grown-ups in this game, kiddo. We've all got to live the lives we've got to live."

Pedro's easy aphorisms disguised a real truth. There were still some very important things Pedro wasn't telling.

SOUND
AND
GRAVITY

9

Oddly enough, it was during this period of Mom's increasingly alcoholic estrangement that I began to experience anything like that "normal childhood" one usually encounters only in books. I grew inured, if not accustomed, to the patent bliss of domesticity. I developed a system of routine chores and scheduled ambitions, marking each day on the calendar as I doled out payments to our landlord and utility franchises, milkman and insurance broker. I took two paper routes. I studied every morning and, every evening, fixed both Mom and myself a perfectly edible meal. Two or three afternoons each week I would go out to what I referred to as my "job" in order to earn money with which to put bread on the table.

As a paperboy, I was kept informed of my client's vacations, and so, on routine afternoons, I would break into carefully preselected homes and take jewelry, portable televisions, cordless phones, microwaves, along with many other alluring household appliances, and transport them downtown on the bus, where I sold them at one of the various pawnshops frequented by gaunt men with loose socks who stood about exposing swollen veins in their necks and foreheads, or glowered at me from behind massy and varnished oak countertops as they inspected my merchandise and contemplated ludicrous sums.

"Ten dollars," they said, eyeing me suspiciously, not concerned with where I got it so much as how little I would take. "It isn't worth my trouble. It isn't worth my time."

"Make it fifteen, then," I replied, chewing my impassive bubble gum. "Maybe it's worth a little of your trouble. Maybe it's worth a little of your precious time."

I even acquired during these days a friend. Rodney was twelve years old, and lived in the corner house with his mother, a rather fragmented and conspicuous woman named Ethel. Rodney was the perfect friend for me, really, and introduced me to a world far more disorderly, I imagined, than my own. Rodney was submissive without obedience, patient without serenity. He had a Stingray bicycle, a rather brutal attitude toward his unfortunate mother (which, I must admit, caused me some uneasy admiration for him, as an aborigine might admire the miracle of a cigarette lighter or a beeping digital watch) and a top-floor bedroom filled with marvelous and very dispensable things.

"Why don't you take this shirt," he might tell me. "These are some pants I grew out of. You never change your clothes, guy. You never wash your hair."

Usually I wasn't listening. I was far too preoccupied with

the room's many bright objects to feel at all self-conscious about my appearance. There were board games: Stratego, Pollyanna, Monopoly, The Game of Life, Battleship and Risk. We constructed monstrous machines with red and white Lego blocks, Erector sets and plastic, prepackaged model kits. Mostly, though, I was thoroughly taken with Rodney's chemistry set, a somewhat corroded metal cabinet box which, unfolded, displayed tidy bottles of strange substances with unfamiliar smells, tastes and textures in them. Some of them, like tannic acid, were labeled with urgent red crosses and warned of deadly dangers which should be investigated only "in the company of adults." The set contained beakers and flasks and test tubes and even a small chemical fire with metal clasps and braces. "This is life's sudden start," I said, the first time I saw it. "This is chemistry." I purchased a loose-leaf notebook and began keeping track of the various chemical mixtures I contrived. Sulphuric acid and nitrous oxide and carbon, zinc and pure rubbing alcohol and long-grain white rice. Then, under what I considered "controlled laboratory conditions," I exposed small animals to them. Bugs, butterflies, lizards and frogs. Sometimes the small animals betrayed no reactions at all. Sometimes, a few hours or a few days later, they died. "Science isn't reason, Rodney," I told him. "Science is pure chance and sudden luck. It's magic, in a way. Chemistry is that unstable and perfectly coordinated music of the fundamental that lives in our skin and our shoes. This is where life achieved its sudden flash, and where time itself will someday rediscover its own timeless regeneration." I contributed tannic acid to the beaker labeled POETIC TROPE #117, thiamine spirit and, from Rodney's mother's kitchen cabinet, vinegar, baking soda, and just a touch of oregano. A thin sudsy foam gathered around the beaker's rim. "We'll seek secrets in the random," I told Rodney. "We'll discover truth in chance's sudden dances."

Rodney, leaning against the table and gazing into the brownish fluid, displayed only that marvelous and half-lidded unconcern for which I always envied him. He wasn't after anything, my friend Rodney. He sincerely didn't care if he lived forever or not.

"What about a booger?" Rodney asked. "What about if we put a booger in it?" Without looking at me, he tapped the beaker's rim with the nail of one of his clean, well-manicured fingers, as if trying to startle into existence whatever soft chemical reactions lay down there in the hidden world of chemistry.

The homes Rodney and I systematically violated that spring were very wary places, hollow, haunted and impercipient, like very old lovers or dying trees. Because I was smallest I always entered first, through basement windows, up trellises into high bedrooms or, more usually, through the opaque and slender windows of bathrooms which had been left open to air out the muggy shower smells. Then I would come around to the front door where Rodney would snap his gum at me with his weary and somewhat affected nonchalance and help me peruse the belongings of these soft and dimly dreaming houses.

"What a bunch of crap," Rodney said. "What are we going to do with all this crap?"

Rodney was an idealist, and refused to be corrupted by mere matter. If I was a sort of exemplary enlightenment scientist, Rodney was a romantic poet, airy and uncompromised. "Crap crap crap crap crap," Rodney said as I loaded pearls and sparkling brooches into my green plastic Hefty bag, watches and piggy banks, digital clocks and compact discs. "They'll never even notice it's gone. They're probably at the shopping plaza right now, buying more crap." He shook his head wearily, and poured himself a stiff drink from the liquor cabinet. If he found a pack of cigarettes on a bedroom bureau or kitchen counter he would chain-smoke casually, filling these transgressed homes with the roiling, misty odor of Marlboros and Kools. I had great hopes for Rodney in those days. I believed then, as I believe now, he was destined for far greater achievements than myself.

"Good riddance," he always said, slamming shut the garage or front door as we walked down the suburban streets with our loot. We wore the purported innocence of childhood wrapped around us like menacing cloaks and fog in some old movie. Only Rodney and I knew what we hid inside those cloaks. Only Rodney and I knew the secrets of the movies we lived inside, the movies other people only watched on TV.

These were the days of my exile, a time of dense silence, strange houses and broken basement windows. They contained locks that could be uncranked with tire irons, or cats that purred and rubbed themselves against you. Sometimes the dogs barked, but if you approached them in a certain way they would bow submissively and allow you to scratch their foreheads. Sometimes we fed the pets while we gathered up the belongings of their masters, and they curled up purring and dreaming on the living room carpets where we would activate the TV for them, for Rodney and I also felt more at home with the sound of the television around us. Game shows filled with jeering buzzers and brand-new cars. Morning chat shows which interviewed interchangeable circus clowns and members of the board of supervisors. Inexhaustible diurnal melodramas in which beautiful men and women lived and loved, hated and died. Then there was only the resinous darkness moving into the houses when we left them. Sometimes we transported our new stuff home in stray shopping carts; sometimes, brazenly, we parked these indemnified carts outside a McDonald's or Burger King while we paused inside for a well-deserved cup of coffee, a sweet roll or fries. I always knew in those days that this was not the world I really belonged to; it was not my mom's world, which both Mom and I had lost, but a world of other moms and dads I would never comprehend. A stony vast plateau without any landmarks or colors on it. A pale cloudless sky in which nothing moved, nothing sounded. You could walk and walk for miles in this world without ever seeing anybody, except of course at night when you were asleep and dreaming about the dense silence, strange houses and broken basement windows. Locks uncranked with tire irons, purring cats and submissive, basement-anxious dogs. Exile was a dream of a return to something you couldn't remember. It took you back to a place you'd never been.

"I think we should burn the dump," Rodney said sometimes, languorously reviewing a *TV Guide* on the living room sofa while I did all the hard work, disengaging the VHS from the Panasonic, stuffing my coat pockets full with quarters from a tin cookie jar in the kitchen. "I think we should see if shit burns." Rodney never seemed the invader of these broken homes, but rather their more legitimate occupant, as if his invisible royal blood admitted him to secret kinships and demesnes. Sometimes I felt rather awkward, looting the silver and jewelry before Rodney's calm and disaffected gaze. It was as if Rodney was allowing my trespass and at any moment, if I made one wrong move or discourteous gesture, such license would be summarily revoked. His expression always seemed remotely curious whenever he looked at me, or at the items in my hands, as if he retained some unflagging interest even though many thousands of years ago he had given up the possibility of ever being surprised again. "There's a good movie on Channel Four we can watch at my house," he said. "It's got Ginger Rogers in it. I think Ginger Rogers is a great piece of ass, don't you?"

All these houses seemed like one house, just as all the

silence of my strained exile seemed like one continent, one forlorn place without a name. I could hear my mom in these houses, I could see her dazed looks as she sat drinking alone in her room, waiting while Dad gathered somewhere in the world like moisture, like thick clouds, like heavy black currents. My sense of exile was my inheritance from Mom; it might somehow, without my even understanding why, constitute my one real gift to Dad, to whom I still owed the ominous debt of conception. I was off in the world alone now. I was investigating strange rooms, basements and gardens. I was trundling off with my pillowcases and Hefty bags filled with merchandise like some sort of diabolical and inverted Santa Claus. All of the houses were part of one house. All of the houses in the world were part of that one house by which Mom and I were divided as well as embraced. "Growing up" began to signify one thing only to my feverish imagination. Mom and I could live in worlds without each other in them.

I never understood Rodney, but I was always a bit awestruck, if only because of the incomprehensible life he lived with his own mother. Ethel had a generous pension from the Marine Corps subsequent to her husband's death at Tet, very gray hair, and bad circulation in her legs. Usually she sat all day and embroidered in a big stuffed chair, her feet propped by cushions and a macramé footstool; when she walked she walked with the aid of an aluminum cane. Whenever we came in the front door with our new stuff she would put down her knitting and watch while we stored it all in the large hall closet alongside the departed Mr. Johansen's crisply dry-cleaned military uniform, unused golf clubs, and loose photographs in a chipped Macy's gift box (I was forever examining the contents of other people's closets). After we were finished, Ethel offered us food and refreshments. "There's tuna salad, Roddy. In case you and your friend are hungry. There are some Snickers bars in the freezer, just the way you like them. Only have some tuna salad first. Have some good canned soup—there's mushroom and tomato. Then, if you and your little friend want, I could fix us all a Manhattan."

Rodney said, "Mmmm." He went into the kitchen and began banging cupboard and refrigerator doors. I stood noncommittally in the hall, watching Ethel in her chair. Ethel was reading one of her old "collector's" editions of *The Amazing Spider Man*, and the plastic envelope lay across her knees like some official procedure. "There's Sara Lee pound cake, and even a couple of Twinkies hidden away. And of course I could fix you both that Manhattan. Would you like a Manhattan, Phillip?" She started to lay her comic on the coffee table and reach for her cane.

"Do me a favor, Ethel," Rodney said. He had suddenly appeared beside me, one foot on the stairs. He held a pair of tuna salad sandwiches on a white plate, and a large bag of Nacho Cheese Flavored Doritos under one arm. "Just sit down, read your comics, and shut the fuck up."

I couldn't look at Ethel. I couldn't look at Rodney. I felt a deep painful turning in my body. My face was filling up with heat. I was walking through a stunned silence, my feet on the stairs, Rodney already at the top. I was still trembling. Everything was a blur. I could hardly see where I was going.

"Don't tell me to shut up," Ethel said, quite simply and unemphatically at first. It was as if she were telling us where the mayonnaise was. "Don't tell your mother to shut up. Rodney. Rodney, come back here."

Breathing a long sigh, Rodney gestured me into his room. He handed me the plate of sandwiches. Then he shut his bedroom door firmly and locked the flimsy knob.

Ethel's voice was growing louder now. "Don't tell your mother to shut up, Rodney. Rodney! Don't you dare tell *me* to shut up! Rodney! You come down here! Rodney! Why don't *you* shut up, Rodney! Why don't *you* shut up, then! *Rodney! You* come down here! *You* shut up, Rodney! *You* shut up!"

Rodney pulled a pair of Cokes from underneath his bed and ripped them free of the stiff plastic spine. "It's like living in a madhouse," he said, not even looking at me. I felt complicit in a frame of violence I couldn't understand. I just sat there hoping his mom wouldn't remember what I looked like. I just hoped Rodney's mom wouldn't remember my name.

"Sometimes I think she's the biggest asshole in the entire universe," Rodney said, pulled the television closer on its wobbly castored frame and switched it on. You could hear the charge of it before you saw the light abrupt to its screen. Suddenly we were in any house, every house; suddenly we were drifting again through the space of my exile. We watched cartoons, movies, detective and western programs while I listened to the outside hallway for the steps of wounded Ethel on the thin carpet as she moved, slowly and eventually, to her own empty bedroom down the hall, awaiting the moment when I could escape this house and my own complicity in Ethel's systematic humiliation by Rodney, the most remarkably powerful person I have ever known in my entire life.

10

I celebrated my birthday in secret that year, on a day I firmly refuse to commemorate or even mention. There was something firm and round about the new age which filled my body like a very old song, or smoke from a cigarette. Usually I didn't return home until one or two a.m., since Rodney and I regularly stayed up drinking Ethel's whiskey or smoking Rodney's grass. My feet staggered and slipped against the knotty carpet as I let myself in the front door. My tongue felt thick and swollen. I staggered down the dark hall, already sensing the thick silence behind Mom's steadfast door. "Mom," I said, leaning against her door, uncertain of the floor's balance. "Mom, it's me. It's your son. Mom. It's Phillip." I heard my own whispered words deep in my throat and chest, resonating like bones. I could hear her taking her breath as my hand gently grasped the loose aluminum doorknob. The knob ticked in its frame when I turned it. Its resistance was at once strange and comforting, like the taste of a new tooth. Mom's door was always locked. She never let me in anymore.

"Mom." I tried to sound firm now. I tried to sound sober and mature. "There's money on the kitchen cabinet. There's still some Colonel Sanders in the fridge. It's cold, Mom. Just the way you like. And coleslaw. Have a banana. Bananas are filled with potassium." Motes and air whirled in Mom's dark room, rustling and indifferent. This was the sound Mom lived. The long slow pause in her heart where she gathered language and waited for history to resume again.

"Thank you, Phillip," she said, as obliquely as she might acknowledge some porter in a hotel. "Thank you very much."

"I'm in my room, Mom. I'm in my room if you need me." I felt the pulse of alcohol in my blood as if the entire house were contracting gently around me. Then I heard the unmistakable gurgle of liquor being poured into Mom's smudged glass.

"You're a very good boy, Phillip. Don't worry about me. I'm fine. I'll be all right. You just make certain you're going to be all right, too." There was a rustle of newspapers. I could feel the darkness assembling in Mom's room, like clouds and gulls around some alien shoreline. For months I thought I was the one who had eliminated the buzzing opposition of Mom's men, but now I knew it was that gathering darkness. It was descending in the elevator from its high luxury office building. It was accepting the keys to its

Triumph from the black attendant. It was flying off across freeways and cities. It loved us. It loved Mom and it loved me. It loved both of us very, very much.

I didn't think of Ethel as a surrogate so much as a compensation. She could never take my mom's place, but she could make that place seem less cold and drafty. Some days I arrived deliberately early at Rodney's house, when I knew he was still in school, and drank generous Manhattans with Ethel while melodramas played at us from her blurry black and white television (the good color television, of course, was in Rodney's room). Eagerly she told me all the lost, distracted secrets of her prodigal son. "Rodney is actually a very affectionate young boy. Like you, Phillip, he is patient and attentive. He's a good listener. He's considerate and well-mannered—when he wants to be, that is. He always helps me with the housework if my legs are sore. Sometimes he saves money from his allowance and buys me little presents. If he's rude, it's just because he likes to show off in front of his friends. Young men, as you know, are a little embarrassed to show affection to their mothers, particularly when their friends are looking."

I wanted to ask her why, but was afraid such a hard awkward question might give me away. I might suddenly divulge the secret life I lived in strange houses, the secret life my mom had begun living behind the bolted door of her minimally furnished room.

"Would you like another Manhattan?" Ethel asked.

She showed me her empty glass and I took it.

"I'll get these," I said, and returned to the kitchen for the Jack Daniel's. Ethel had taught me how to mix a number of competent drinks, a feat in which I admit I took some pride. I returned and sat on the sofa beside the morning's smeary newspaper and watched the fabulous television. Ethel was absently handling her embroidery frame and gazing out the window at the harsh, smoggy sunlight, the palm trees faded and unraveled like some overexposed snapshot, the uniform houses and pavements and flashing cars. "People don't always intend to make other people feel bad, Ethel," I told her, though I was never sure she was listening. "Sometimes people just forget other people are even around. I know it sounds strange, but people work that way, I swear. Sometimes they don't know what they're doing. Sometimes they don't even know you're there at all." I was grasping at straws. Whenever I found myself trying to excuse Rodney's disgraceful behavior I became tangled and caught in my own inflexible words. Ethel, meanwhile, gazed out the window. "Maybe people just don't know where they are sometimes," I said, afraid to stop talking because then the judgment would come. In the long pause my talk would have to mean something. "Maybe people just talk without remembering who it is they're actually talking to. Maybe you just shouldn't think about it, Ethel. Maybe you should join a health spa, or find a hobby that interests you. Do you hear me, Ethel? Would you like another drink? Ethel? Are you listening?"

Before too long I was taking lunch with Ethel every after-noon around one o'clock. The casual scheme of my domes-ticity was growing more fulfilled and content. My paper route, breakfast, morning study sessions (I was currently investigating Plato, biophysics and Freud), afternoons with Rodney burgling strange homes, television, evening meals and bed. And every afternoon before I left the house I would leave Mom's lunch wrapped in plastic and deposited outside her bedroom door. Ethel was instructing me in the art of fine sandwich building. Tuna and chicken salad, avocado and sprouts, bacon, lettuce and tomato, roast beef, pastrami and turkey with cottage cheese, peanut butter and banana. Whenever Ethel sliced the sandwiches in their rich brown bread the divided segments always looked impossibly tidy and controlled on their clean white plates. "Sometimes at night," Ethel said, humming and staring out the back win-dow while she washed her hands at the sink, "sometimes at night Rodney's father calls me from a long way away. He says he would like to move back in here. He says he misses my cooking." Ethel's voice trailed off aimlessly, like late night drivers descending off-ramps in search of a quick, in-expensive meal. "I tell him I wouldn't mind, if it was just up to me." Drying her hands on a thin patterned towel, Ethel was gazing over my shoulder. Her eyes looked so intent, I often turned to see what she was seeing. I suspected a grown man had appeared behind me, perhaps a taller and more mature version of myself, Ethel's more substantial companion which my thin body merely represented. " 'But

80

it's not just up to me, Harold,' I try to tell him." Ethel was sculpting soft white flower petals from the bodies of scoured radishes with a small sharp paring knife. " 'We have to do what's best for Rodney. We have to do what's best for our son, who has been raised under very trying and unfortunate circumstances, as I think you well know. It's easy, being a father, to think you can just show up when you feel like it. But children need someone they can count on, Harold. And as much as I may want you back, I just don't think Rodney could ever count on you again. You'd only disappoint him.' "

Ethel never disturbed or embarrassed me. I knew she had her own secret life to live, just as all mothers live fair portions of their lives down there in dark secure rooms and hidden gardens filled with strange plants and trees. I was simply grateful for the time Ethel spent with me here on the outside while I learned to prepare bases for soups and gravy, toss Caesar and fruit salads, cook purées and stews. I fricasseed, baked, boiled and roasted. I cleaned chicken and fish, basted lamb and pork, pressed my hungry hands into the thick dough of breads and cakes and cookies and pastry. I loved Ethel's warm kitchen and the heady smell of bread baking while I waited for Rodney to return home from school and transgress with me those other, colder kitchens, where I was picking up handy appliances, kitchen pots, pans and utensils and reassembling them in the hard irrefutable kitchen of Mom's silent and discriminating house. I wanted to build Mom a strong home that would always be there for her and provide her anything she might ever

need. I was beginning to realize I would have to leave someday. I still loved Mom more than ever, but I was learning
life carries us places, like rivers and winds carry things, often
against our own will.

Rodney may have been only twelve years old, but he had
big ambitions. "I need real estate," Rodney often said after
his second or third drink. With our merchandise gathered
around us in the living room like a family at Christmas, we
were lingering overlong in a particularly commodious four-
bedroom Spanish-style home in the foothills of Sherman
Oaks. Outside the streets were filled with dry, amber lawns
and stark, shedding palm trees. Every once in a while a
bright bluebird flashed. "I need tax incentives, money market liquid asset accounts, diversified stock portfolios, trea-
sury bills, low-interest tax deductible loans, property, houses,
income property, cars, trucks, buses, planes. I don't need
this. I don't need this crap," he said, making his customary
gesture at the huddling portable televisions, radios, jewelry
and microwave and, still in its original Sears packing case,
an adjustable three-temperature electric blanket which I in-
tended to leave that night, like a meal or some religious
devotion, outside my mom's bedroom door. "I don't need
a bunch of crap just weighing me down. I need negotiable
capital. I need security and a firm financial investment base.
I need money, property and women. I'm talking gash, now.
I'm talking poontang. I'm talking scuzz. I'm talking count-
less good-looking, insatiable young women with big tits."
Aimlessly Rodney's left hand began stroking the inner thigh

of his Levi's. Then, abruptly, he leaned forward and reached for the margarita mix in the blender's thick Pyrex bowl. "I need money and sex and more sex. I been thinking about it every day lately, Phil. I gotta get laid, man. I really gotta get laid."

Then, contemplating his refreshed glass for a moment, Rodney slumped back against the sofa. Meanwhile, I tried to appear as calm and unaffected as a prayer. I was practicing with a cigarette, sucking the thick smoke into my mouth every few seconds and then expelling it, pushing it out with my tongue and cheeks. Phooh, I said, as quietly as I could, because it wasn't a sound Rodney or Mom ever made when they smoked. Phooh. A very ugly miniature terrier lay asleep dribbling in my lap. Gently I lifted it onto the sofa's side. "Don't talk about it," I told him. "Don't talk about what you want too much. You'll lose the edge."

"I need to fuck women." Rodney's voice was growing subdued, distant, ritual and dark. "I need to fuck fuck fuck until I can't fuck anymore."

"It's all a dream, Rodney. If you talk about it too much, you wake up. Then there's just the bright sun. Then there's just your cold bed."

Suddenly Rodney sat up straight and placed his margarita on the table. He cautioned me with his left hand and gazed off intently at the far wall, as if listening with his eyes, poised like a diver.

I heard the footsteps too. Keys being shaken. Then a sack of something banging against the porch while keys rattled more distinctly. Implicitly feminine sounds.

83

"If you keep dreaming, you can have it all, Rodney. If you keep dreaming you can even be a grown-up. You can even fall in love."

"Fuck love," Rodney said.

We were grabbing the most compact and obvious loot, then slipping down the back stairs and out the rear garage door, through the yard and backyard gate while upstairs that stupid terrier was yelping and throwing itself up and down in the air as its master incautiously opened the front door.

"Fuck love, man," Rodney said later at Burrito King. "Fuck family. Fuck people and things. I want the real *stuff*, man. I want currency, I want sex. Give me the *stuff* and save all the bullshit. I've had it up to here with all the bullshit, Phillip. I'm tired of rotting away in Ethel's lousy household. It's time I started living my own life. I'm telling you, guy. Life is something you do. Bullshit is just something you've got."

Currency and sex were forces in our lives now, like smoky, violet surges of electricity and light. Sex and currency, currency and sex. The hum and the pop, chirring and turning, beating like electricity. We could drive cars with that force radiating deep inside us. We could activate industrial machinery. We could generate enough massive interior energy to drive cities, planets and suns.

I didn't feel the same sudden push inside my flesh that Rodney felt, but rather a sort of anxious intellectual charge. The energy whirled aimlessly inside my mind, where I nightly replayed that sudden film I had viewed with Mom

84

only a few months before on a hotel-room closed-circuit television, *Sexually Altered States*. Sexually altered, sexually altered altered states, states of sexually altered states, altered states of sexually altered states. Sometimes the images sped and raced in my mind, and I imagined myself taking a seat in my own subjective cinema. I imagined the curtains sweeping open and the light dimming. I imagined the credits, and then the first exuberant breathless cinematic fuck. I tried to contain all the events within the frame of a plot. I tried to imagine the interstitial scenes, the dinners and champagne, the slow dances and undress, often growing so involved in them that I never got around to thinking through to the serious action again, the way the karate master is supposed to think through boards and blocks of concrete. I thought it was the anticipation which made sex real, but now I know it was merely my explicit faith in that imagination which unreeled around me in my childhood like the spokes of a milky galaxy. I was beginning to learn that the imaginative act was more important in my life than action itself. Action merely articulated you with an exterior and superficial network of facts, data and information which superimposed itself across the real world of my imagination like a restraint, or a clinging, oily film. I wanted the dream of sex, the energy and heat of it. Had I been able to articulate the problem for Rodney, I'm certain he would have preferred the dreams too.

11

"I've never considered myself beautiful," Beatrice said, listlessly running her hands through her tangled blond hair. Beatrice was twelve years old and attended Rodney's sixth-grade class at Junípero Serra Grammar School. "But I don't think beauty's the only thing men are looking for, if you know what I mean." Her look let us know she didn't expect there was any chance we would ever know what she meant. She tugged at her slightly soiled dress and crossed her legs. After a second of demure hesitation, she accepted Rodney's cigarette. Then Beatrice lit it with a Ronco from her purse, which she snapped shut with a practiced flourish. "Beauty's just what a woman seems. Plenty of women can be made beautiful. Look at *Vogue*, for instance. Look at *Cosmo*. That's not all men are looking for, you know. I *know* that's what they say men are supposed to be looking for, but that's just a male myth. That's just capitalism. That's just the psychological domination manifest in all competitive class struggle.

Basically, you see, I think men are a lot more capable and intelligent than that. I think men want a woman they're attracted to, sure—I'm no spiritual idealist or anything. I don't subscribe to bogus Christian dualism—all that repressive male ideology we've inherited from the Greeks. But there's a natural woman men are looking for underneath all the Clairol and Maybelline. There's a raw—oh, I don't know, call it sexuality, or passion, or molecular urgency—that marks a woman out from the pack. It's in her eyes, in her smell, in the way she combs her hair." Delicately, Beatrice jiggled in her seat, tugging her skirt straight underneath her.

We were all sitting at the Formica booth in Winchell's, feeling that warm arousal of steam from our fresh coffees. I held my Styrofoam cup between my hands. I was trying to keep my eyes expressionlessly focused on Beatrice. I was certain any expression on my part would be a sort of self-betrayal. Better to give her nothing about me she could analyze or remember, nothing she could keep for years and years, like nail clippings or stray buttons with which she could cast intricate social spells. I just wanted to watch Beatrice, her lips so hastily smudged with the chocolate rainbow donut, her pink skin and knotty, unwashed hair. Beatrice lived in a trailer park with her father. The trailer park, located in Encino, was called Trailer Town, and Beatrice's one-bedroom trailer was a 1959 Spartan Luxuryliner with polished wood interior and fully operable stove and central heating.

"I think that's what I'll always have to offer my men,"

Beatrice said, and took the first brave sip of our scalding coffee. Rodney and I had ordered it black because Beatrice had. She put it down with a little emphasis.

"It's like television, movies, books, even record albums. *This* is what beauty is. *This* is how you're supposed to look. *This* is a girl's *normal* height, how much makeup to wear, how big your tits should be"—I felt Rodney give a little jump beside me—"and all that long morose unforgiving catalog of what women are *supposed* to be, how women are *supposed* to feel. Beauty is the culture industry's attempt to make each of us a commodity. The culture industry, guys, is vast and incredibly articulate. It knows exactly what it wants to say all the time. It wants to make each of us the same on the outside, while letting us pretend we're somehow marvelously special on the inside. The culture industry hasn't invented 'beauty' in order to control how we look, but how we *are*, and that's the scary part. How we think. How we *be*. I guess you guys should know right away I'm a Marxist. I support the Sandinistas, and the leftist guerrilla forces in Chad. I'm not a vulgar Marxist or anything—I mean, I don't pick my nose in public (that's supposed to be a little joke)—no, I guess you'd have to call me a post-structural Marxist. I give credit to Althusser, but I'm not an acolyte of *anybody's*. Anyway"—with another little flourish of her black purse—"if you'll excuse me, I've got to find a ladies' room." Beatrice glanced over her shoulder, then pointed. One tiny cuticle of her index finger was perceptibly tagged with chocolate. "That's it there, I think."

When she was gone, both Rodney and I took long, contemplative sips of our hot, bitter coffee. Around us the scrubbed linoleum still smelled strongly of ammonia and disinfectant, and every once in a while the matronly woman behind the counter, wearing a black hairnet and plastic gloves, gave us a rather dirty look, as if she expected us at any moment to run off with something. Coolness drifted through me. I was very high in the air, drifting among birds and planes. It was incredibly quiet in the high air. There were no words in sight, not even fragments of words. Eventually, without looking up, I leaned toward Rodney.

"What's she talking about?" I asked.

Rodney didn't say anything right away. I heard the flash of a match which he dropped into the black ashtray. "I don't know," he said after a while, watching the bright match flare, beat and extinguish with a tiny puff. "But I think it may be another load of crap."

We would take Beatrice to my house in the afternoons when we weren't scavenging other houses, but she never put out. She always said she was going to put out, but she never really did. Though my interest in the outcome was philosophical rather than immediate, I would watch the slow struggles Rodney waged with Beatrice on the poor, makeshift couch in my living room with perfect equanimity. They would kiss for hours, penetrating into the roots of one another's mouths, breathing deep into one another's lungs and hearts, shifting and turning very slowly, Rodney's leg between Beatrice's, his arm around her waist. Sometimes I would watch them for a while, but then I would watch the television instead. "The Rockford Files," perhaps, which I adored, or "Barnaby Jones," which I could at least endure. "Mmm," Rodney might say, though Beatrice herself always remained chastely, even demurely silent. "Mmm, baby." It didn't sound quite right to me, but perhaps, I thought, Rodney was still practicing. Perhaps love wasn't something you felt, but rather something you learned. Then, abruptly, Rodney's hand would stray too far and, before either of us knew it, "it" would be abruptly over.

"It's just not right," Beatrice said, sitting up and running her hand through her stringy hair. One leg was folded underneath her, and a hard, reasonless and abstract gaze made her eyes seem very cold and distant. "I guess I'm not in the right mood or something."

"What do you mean it's not *right.*" Rodney was running his hands anxiously through his own unsprung hair. He resembled a broker during a crash, his shirt rumpled and undone, his eyes slightly wild and bloodshot. "What's *right* like? How am I supposed to know the *right* time? How'm I supposed to know what the *right* mood's like?"

Sometimes she petted and tried to console him, and sometimes they would even start kissing again, their thin lips bumping again and again at one another. Then their bodies would do that slow horizontal dance again, and I would watch just for that movement, that steady and directionless rhythm of their bodies on the sofa. Bits of foam rubber were spilling from underneath the tangled bedspread like sawdust from a lathe. My interest was merely clinical.

I knew it was the vital dance. I knew it was something very serious and inexplicable for Rodney and that, with luck, it would someday be the same for me as well. But for now the irresolvable dialectic of that motion was what fascinated me. I wanted them to be like that forever, like a glittery mobile or a perpetual object of performance art. I wanted them moving there together on my sofa until I forgot about them, until I accepted that movement of theirs as calmly as I accepted the walls and ceiling of my house, as firmly as I accepted my mom's secret and brooding presence in the silent back bedroom. Motion not as a way of living, or as a dance of bodies, but as a sort of universal presumption. Bodies moved, cars moved, planes moved in the sky. Rocks and trees and garbage cans and concrete cinder blocks moved. The earth moved and the stars moved. There was the spiral movement of entire galaxies, and the fundamental movement of atoms and quarks and merely theoretical matter. Everything was moving everywhere all the time, because nobody was anywhere they wanted to be in the first place. Time, when you considered the elemental, even archetypal fact of motion itself, was just a formality, a record, a graph. Movement was life, and when you moved your body into the body of a woman you initiated other movements, other lives. Cells moved, proteins and enzymes moved together, particles of minerals and plasma moved in the blood. I didn't care whether Rodney consummated his obsession with Beatrice or not, because I had come to realize that motion itself was destiny enough. Its consummation would only inaugurate a moment of dulled, sleepy disavowal. You could pretend motion had stopped for a little while. Then you would lie down and sleep until your dreams woke you.

"You know what America is, don't you, guys?" Beatrice asked us one day, examining herself in her compact, plucking at one damp, tangled eyelash. "It's a big black hole that sucks everything in. You know who founded America, don't you? People who could pretend they were anybody they wanted to be, because that's what America is. Anything you want it to be."

"America's the frontier that's never conquered," I said, reaching to turn the television volume down. I swiveled around on the rug to face her, feeling something slip in my stomach, my groin, a feeling as if I had swallowed something very cold and heavy. "America's motion, America's always somewhere else. If we can't go other places, we can be those other places instead. My mom always said that's why we're Californians—because we don't need to be here. We can always be anywhere else in the universe besides California, and still be here too. America's the dialectic. It's what Hegel talks about in *The Phenomenology of Mind.* The dialectic."

Somewhere deep in my house I could hear Mom silently nodding her head. She wasn't approving, though. She was just nodding her head.

Rodney let out an exasperated sigh, and reached for his Dos Equis. "Jesus Christ. Just what I need." He took a long pull from the beer and quickly lit a Tareyton. He looked dispassionately from me to Beatrice, from Beatrice to me.

His face was puffy and creased with worried lines. "Fucking stereo," he said.

I felt very good sitting there, watching Rodney and Beatrice on the couch. Rodney looked at his cigarette. Beatrice, out of the corner of her eye, looked at Rodney.

Life had grown very substantial and real for me while my mom lay quietly in her room. I had everything now. A family, a house, a lucrative job, good times and faithful friends. I had the sweet clutter of books in my room. I had cigarettes, good whiskey, excellent home-cooked meals, a reputable broker, new shoes, the privacy of my own mind, healthy sexual curiosity and, somewhere in the world, Dad, who would always take me back if things got too rough. I felt settled, but not conventional. I was learning how to live my own life and yet still love Mom too, the lesson I knew Mom had always meant for me to learn. Mom was very real and immanent all the time. She was a vast incontrovertible force, extensive like gravity or sound. She was like God, she was like air. And she was always in perfect control, especially when she wasn't in any control at all.

1 2

"Is your mom seeing anyone?" Dad asked me, over and over again. "You know you can tell me. You know you can tell me if she's seeing someone." His voice grew somewhat webbed and anxious. There was something sudden about Dad's voice when he considered Mom's possible infidelity. He reminded me of Rodney, exasperated by Beatrice's muggy sex. He was always very quick to change the subject. "Is she all right?" he asked. "Is she eating properly? Do you have enough school clothes and spending money? Should I wire you some? Should I stop calling? Should I let you go on pretending I don't exist at all?" Dad's voice strained against the force of our lives, divided from us by a thin, translucent bubble. The bubble's skin transmitted Dad's voice and texture, but not his body, not his acting presence. "Does she ever ask about me, or wonder how I am? Do you think some nights she misses me, or says my name out loud?"

"She's not seeing anyone right now, Dad," I said. "She's working very hard, and likes her job very much. She's trying to work out a lot of things right now, and just needs to be left alone for a while. She asks about you all the time, and I tell her that you've been asking about her. She wants you to know she still misses you, but that she needs this time to think about nobody but herself. She needs to live as if nobody else in the world exists except her. She needs to be left alone awhile longer, Dad. She needs to know we love her enough to trust her to be by herself."

Dad's voice was growing more tinny and desperate as Mom's voice, sinking into the silence of her own room, grew more inaudible and self-assured. Sometimes static swirled and seethed on Dad's telephone line like foam on a seashore, leaving behind broken bits of shell and rock and bone. "I think it's time you both came home," Dad said one night. "I think I've been patient long enough. I think it's about time you both learned a little something about responsibility. I'm talking about things like right and moral duty. I don't mean I believe in God, but I do believe in responsibility. And so will you someday, sport. So will you someday."

"What does your dad do?" Beatrice asked one night after I hung up and returned to the living room. The distant, stellar noise of the telephone continued to wheel around me like lights in a planetarium. "Is your dad in business? I always imagined your dad a big, successful businessman, Phillip. It's in your blood, I can tell. You'll be a businessman too, I'm sure. When you grow up."

Dad was still there in the house with us. He was more

idea than thing, more impression than voice. Dad wasn't life. Dad was history.

"It's nothing to be ashamed of—business. Business convenes secret and ghostly ceremonies in the world. Ceremonies which the world needs, or else the world wouldn't have them." Beatrice was gazing abstractedly at the living room wall, turning one long tangled coil of oily hair behind her ear. "It's as real as rocks, as organic as trees." It was almost eleven o'clock and Rodney had gone into my bedroom to sleep. Beatrice held a half-finished glass of Nestlé's Quik in her lap, its rim whorled with milky deposits and fingerprints like the etched fossils of intricate trilobites. "Business always works, even when nature doesn't. When plants stop photosynthesizing, business will manufacture its own atmosphere. When other moons and planets crash into our seas and wreck the world, business will mine and redistribute them. Business is the world's real nature, Phillip. We are all fleeing nature and using what it has taught us about business to make our world vaster and more perfect."

Beatrice had breasts—little breasts, granted, but breasts nonetheless. When I lifted up her mascara-smudged Lacoste T-shirt and reached underneath to touch one, it was as warm as I expected. I brushed it gently with the palm of my untrained hand and then, softly and with clinical care, palpated it the same way a doctor might examine for cysts. I could feel the clusters of glands like tiny grapes all joined up to the nipple's tuberous root. The surface skin was soft. I could smell Beatrice's slightly unwashed odor, like the smell Mom's laundry used to make when soaking in some motel sink. Holding Beatrice's breast in my hand made me feel a sort of cool and intellectual redemption. I don't mean I wanted to be a baby again; I didn't want to be nurtured by this breast. I wanted only to regain a sort of molecular integrity. I wanted to crawl back into the cellular warmth of my own body, not some woman's womb. I wanted to grow so small I could see protons and electrons bristling in my tiny night sky like showers of meteors and cosmic dust. I did not want to procreate that first time I touched a girl's breast, I wanted to uncreate. I wanted to penetrate life in search of the unliving. I wanted to exonerate the fundamental and fragmentary lifelessness of things. Without a second thought, I leaned forward on my tiptoes and kissed my Beatrice on the lips, which were still slightly smudged with Nestlé's Quik. Her lips didn't move a muscle. I don't believe either of us felt a thing.

"You can't hide your desire for me," I said.

Beatrice snapped her gum.

"Our two bodies are meant to be one," I said. "You know it's true. You knew it was true when we first met."

I stepped back, removing my hand from her. I expected I would touch a girl's breast again someday, and remembered the cocoon I once discovered on the branch of a rosebush. I had broken off the branch and placed the cocoon inside a glass jar, the lid of which I punctured with a sharp knife to allow air so the cocoon could breathe. I included branches of other trees and bushes, leaves, stones and dirt for both scenery and a sense of environmental continuity. The cocoon just hung there, day after day, motionless, cobwebby and dry as a bone. Then one day I decided it must be thirsty, and poured in a generous helping of Coca-Cola through the punctured aluminum lid. A few days later the cocoon withered and collapsed on the branch like overripe fruit. Then, one night, when I was asleep, a few green drops oozed out.

I took Beatrice's hand. It was, as always, grimy and smudged. It left a soft glimmering residue wherever it touched.

"I think it's time you met my mom," I told her. Then I took her down the long dark hallway to Mom's silent room.

"Hello, Mom. It's me. It's Phillip. You remember me. Your son."

The darkness seethed behind the door. Dead planets moved there. Somehow comprehensible alien languages whispered and transformed themselves into things. Light wrapped textures around itself, like young children in their parents' clothing.

"I know I said I'd leave you alone for a while, Mom. I'm not trying to hurt you, or break any promises. Dad called and I only told him what I knew you wanted him to hear. There's a fresh salad in the Tupperware bowl on the bottom shelf of the fridge. There's a frozen lasagne you can heat up in the microwave. Now, Beatrice and I will go back into the living room and leave you alone, and if you want to go fix yourself something to eat you can. I promise we won't get in your way, I just wanted you to meet my friend Beatrice. I'm in love with Beatrice, Mom, but it's a completely different type of love from the way I love you. I just wanted you to know."

I could feel Beatrice's body standing beside me in the dark hall. Tonight her body was the only warmth in our house.

"It's love without passion, Mom. I don't feel any passion for Beatrice, just love. She might as well be a rock, or a landmark, or a memory, or a curious bug as far as physical passion's concerned. She might as well be chemistry or math. Anyway, Mom. This is Beatrice."

I gave Beatrice's wrist a little squeeze.

"Hello, Mrs. Davis," Beatrice said after a while.

Beatrice's calm voice resounded in our still house like a summons.

"Now, Mom, we'll leave you alone."

Both Beatrice and I thought we could hear Mom breathing in her bed as we returned back down the dark hall to the living room sofa again.

That night while Beatrice and I slept huddled together on the living room couch I awoke and heard movements in the kitchen. The weak filmy light from the streets filtered through our dusty venetian blinds, thinning and thickening like pages being turned back and forth in a book. I heard a single glass fall and break. I heard rummaging in the ice-maker. I wondered if Mom was feeling OK, and thought I should start leaving vitamins out for her. I heard the coffee grinder, and waxy cardboard being torn from the frozen lasagne. Dishes and silverware began to clatter, the refrigerator door opened twice. Then the initiating chime of the microwave, and I did not hear so much as feel that long slow charge of heatless energy building, a kind of cosmic affinity being conceded deep down there among our secondhand kitchen appliances. The coffee was percolating on the countertop, the aluminum lid rattling. The microwave chimed again and buzzed, and a firm quick hand clicked it off. I could smell the antiseptic prepackaged smell of it, the thawed meat and cheese steaming. Beatrice, asleep against my thin pale chest, stirred; she muttered something. One of her bare legs slowly wrapped itself around mine and she began faintly snoring. I heard the footsteps in the kitchen going back and forth, and boiling coffee being poured into a ceramic mug.

"I think it's time you came out again," I whispered, watching the door to the kitchen which adjoined the living room. "I'm getting worried about you. I don't know how much more neglect I can promise. I miss you, Dad misses you. We all miss you, even the vast world. I'm sure it will all get better. I'm sure if you come out and try to face things, things will get better again on their own."

The noises stilled in the kitchen. There was a slow breathing hush filling up the flashing lacquered walls. There was thinking going on in there. A very strong and willful mind at work. Then, succinctly, a long sipping of coffee.

Beatrice, in her sleep, gently kissed my shoulder.

The footsteps swung through the kitchen again and then, in the doorway, I saw a tall shadow which only slowly, holding the coffee in one hand and the lasagne on a plate in the other, emerged into the dim mothy light of the living room windows.

"What you think you're doing?" Rodney asked, wincing against the bitter coffee. He gestured with the cup at Beatrice. "That's my woman you got there." His face was puffy and his hair stuck out like Larry's on "The Three Stooges." He stood there for a few moments, not looking at Beatrice and me so much as through us. Then he sat down on the floor and began to eat his lasagne. His fork ticked dully again and again against the cheesy plate.

"Sometimes I don't know," Rodney said, wiping his brow with the back of his greasy hand. "Sometimes I don't know if I was born mean, or if the world just made me that way." He lay the dish and fork on the floor with a deliberate clatter. He looked at the musty, glowing venetian blinds. Outside the phototropic streetlamps were beginning to dim; the power lines and parked metal cars began to buzz slightly and deeply on the empty street, filling up with the sun's early warmth.

"I think Dan Duryea said that," Rodney said. His sleepy eyes remained trained upon the cracked venetian blinds. "I really get a kick out of that goddamn Dan Duryea."

13

The following evening we finally decided to take Mom's car out for an exploratory little spin. The engine started rough, but it started, and I let it warm up in the strange and generally unglimpsed garage of Mom's house where everything, even the shadows, seemed inverted and unreal.

"We got wheels!" Rodney declared, jiggling in the front passenger seat and thrumming Beatrice's thighs with a paradiddle, a flam-paradiddle, then a spotty double roll. "Phillip's got us wheels, man! Let's *roll!* Let's do like they do on TV and let's *roll!*" It was the only time I ever saw Rodney excited about anything. Rodney even offered to give me five, but my mind was too intent on the various pedals and knobs and levers before me. They were all familiar, but they were all unfamiliar too. They articulated me with that world of adults which was accessible in every way except the mind's. When I touched them, I was touching some ineffable something of adults which made adults different. It was not the things themselves, but some mysterious identity behind the things. Signals, ashtray, wipers, defrost, tone and volume, fan, air, seat belt, dash. I eased the accelerator down and up, the engine expanding in the damp garage with a choppy roar and then diminishing again like passing private airplanes. I practiced engaging the clutch and killed the engine three or four times. Each time the car started with a throatier, smoother voice. It was the voice of mechanical discretion. It was the voice of the whirring conspiratorial notions of engaged mechanical parts.

So that was the winter I learned to drive, propped up by two volumes of the Yellow Pages, and always at night. It was a dry, brittle winter in the San Fernando Valley, and the only snow around was painted on the shopfront windows of barbershops and Exxon stations. Santa Claus, flanked by bikinied elves, sweated in his bright red underdrawers as he lugged his gifts past high palm trees and a bright yellow sun. The streets were filled with young girls and boys standing about, leaning against their bicycles and eating ice creams and candies and bright orangish Slurpees they had just purchased from 7-Eleven. The air was heavy still with the always rich jacarandas, as if winter never came.

Some nights it rained, and we drove our car through long ripping puddles in the flat streets of the poorly irrigated basin, around Valley College, down Van Nuys and up into the Burbank Hills into Glendale, Pasadena, Whittier. When it rained the entire San Fernando Valley flooded with the bright rainwater and the reflected lights of other cars. I liked to downshift at stoplights and then, in neutral, coast through when no cops were watching and jerk quickly into second again, sometimes feeling that sudden loss of gravity when the tires spun and the smoke and water steamed together and wrapped us up in a world very invisible, yet also very real. The windows of the houses all betrayed bright Christmas trees with silver tinsel, red bulbous ornaments and inconstant strings of flashing lights. Enormous plastic Santas, reindeer and manger scenes stood out on the front yards like migrant workers, and whenever we turned on the radio they were playing Christmas carols. "Jingle Bell Rock." "A California Christmas." "O Come All Ye Faithful." "O Little Town of Bethlehem." O Christmas, I thought. O Mom, O Dad. O History and Motion, O Motion and Light.

"Turn off that crap," Rodney said, still happily thrumming his hands—against his own thighs now. Another of the bitter silences had interrupted his necking with Beatrice, who sat between us in the front seat now, one hand surreptitiously on my knee. She was sucking a LifeSaver. "It's still fucking November," Rodney said. "I haven't even digested my Thanksgiving fucking turkey yet."

"Do you believe in Christ?" Beatrice asked, gently disengaging Rodney's hand from her shoulder.

"You mean like Christ, the son of God?" I had just lit my Tareyton, and was reinserting Mom's lighter into the dash.

Beatrice, her eyes somewhat dazed by the glimmering streets, nodded.

"You mean like the bread and fishes?" I asked. "You

104

mean like the Star of Bethlehem, and the three wise men, and dying for all our sins?" I was starting to get a little excited. "Christ wore a halo on his head, even when he was a baby. He slept one night with the carpenter. He gave speeches on a mountaintop and collected apostles like trading stamps. One day they nailed his hands to a cross. Then, a few days later, he lived again. Which always made me wonder why God let him die in the first place. Because it was a symbol, that's my guess. God never cares about human beings. God, like any halfway decent politician, only cares about symbolic language."

"I dig the story about the bread and fishes," Rodney said. "We're talking luncheon meat for forty thousand. We're talking pimento, and coleslaw, and fried chicken. We're not talking your average lousy tuna salad. If my stupid mom had been there, she'd have been asking why they didn't serve any fucking tuna salad."

"I think Christ is an idea we came up with because we needed it," Beatrice said. I glanced at her in the rearview mirror. Her eyes were still closed.

"I think we might as well believe in Christ as believe in anything," Beatrice said. "I mean, Christ makes just about as much sense as anything else I can think of."

"Like Success," Rodney said.

"Or Self," I said.

"Or Family," Beatrice said. "Or Woman. Love. Disease. Heartbreak. Death. God. Goals. Reification. Fried food. High fiber optics. Disinvestment. Cancer. AIDS. Genes, skin, tissue, soul." Beatrice's face and neck were covered with so many of Rodney's red, splotchy hickeys she resembled a victim of the Great Plague. "They're all ideas we need. The question shouldn't be whether they're real or not, or whether we believe they're real. I mean, I can't tell you how impossibly *mundane* that all is, all these ridiculous endless arguments about what's reality and all. They're such imbecile restraints upon our thinking. Empiricism isn't a way of thinking; empiricism is a way of being. Of being a dickhead, anyway. Do we *believe* we need the idea, that's what interests me. Do we *believe* we need Christ, Phillip? Or do we believe we have any good reasons for *not* believing?"

"Don't forget childhood," I said. "Childhood's one thing don't forget."

"I believe we got to stop at a service station so I can drop a log," Rodney said. "I mean, I'd really *love* to continue this highly intellectual conversation and all, but first I think—I mean, I *believe*—I *believe* I've got to chop me some wood. I *believe* I've got to cut me one mighty humongous loaf."

Whenever I returned home Mom left traces of herself around the house just so I would know she was all right. It was like a secret Morse of displaced objects, punctuated by unvoiced sighs and iconic, invisible gestures. An unwashed glass on the kitchen countertop. A bottle of Seagram's discarded in the trash basket under the sink. Sometimes she would leave a pillowcase stuffed with dirty linen outside her door for me to wash. Sometimes there might even be a little note attached to it. Usually it said:

Dear Phillip,

I love you.

Love, Mom

And of course I always left Mom a note in return:

Dear Mom,

I love you too.

Love, Phillip.

"You can't grow up thinking life's like your mom lives it, Phillip. I think that's the important thing. I think that's the real reason I want you back home with me. I'm not trying to say it might not seem fun, especially to a young boy. But what seems fun isn't always right, and I think you know what I'm saying. What's right isn't always fun. I'm talking of course about all those corny, old-fashioned values your mom used to ridicule me for trying to hold on to. Things like honor, commitment, duty, and yes, even good old-fashioned family loyalty. Hell, I'm not trying to sound like some sentimental old fool or anything. It's just that the love you feel for your lawful wife and child happens to be a lot more 'fun' in its very difficult, thankless way than any free ride in your mom's car. I think you know what I'm saying, Phillip. If I'm not too far wrong about my own son, I think I can say you've got a good idea what your old dad's talking about."

Often after Dad hung up my head continued ringing with his voice. His voice was something I carried with me now, just as I'm sure Dad carried my baby picture in his wallet. "What if everyone behaved like you and your mom. What if the police, or the firemen, or the president just did whatever they damned felt like it, whenever they damn felt like it." That Christmas Rodney and I would park Mom's Rambler in the driveways of the homes we looted. We not only took resalable commodities now, we took the heavier and more flamboyant presents from under strange Christmas trees. We took many of the brighter trees themselves, and lined them up, already decorated, in the living room of my mom's house. "What kind of life would that be like, Phillip, where people completely forgot the responsibilities they had towards other people? I think it would be a pretty frightening world, don't you? What if there was a fire and the firemen were all off having a good time with their friends?" We took festive wreaths and strings of popcorn and hand-painted porcelain angels and music boxes in the shapes of coned trees or leering Santas or elves in workshops. "What if Russia invaded and our president was traipsing off somewhere and hadn't even left a forwarding address? What do you think those Russians would do? I think I can tell you what they'd do, Phillip. They'd walk all over us. They'd walk all over this great land of ours." We found a dog tied

108

to a building support post in the basement of one house, ringed by its own puddled urine and moldering, cocoon-like feces, its skin scraped and chafed under its collar, its leash entangled by its tattered smelly blanket. "They'd start putting people into concentration camps. They'd ruin our industries and entire free market system. The government would tell everybody what everything costs, and exactly how much they could buy. Nobody'd have one ounce of individuality left anymore. Nobody'd ever remember, after a few years or so, how wonderful things had been, and what a terrible mess we made of our great country."

We unleashed the dog and carried it down to our car. It was a dachshund, and we named it Contrite, because it always looked very apologetic about everything. Contrite slept in Rodney's lap all the way home. Sometimes I couldn't hear anything during the day except Dad's voice. Sometimes, though, if I tried hard enough, I could hear Rodney's.

"You know what Christmas means? It means if I love you, I'm going to buy you a whole load of crap. The more I love you, the bigger the load gets. Sometimes, if you can afford it, you can get your loved one literally tons of crap, and then they're really loved. Love love love. It's a terrific idea. Low overhead. Unbelievable mark-up value. I think love is one thing that will always sell really well. Ho ho ho. Merry Christmas." Rodney was rolling a loose joint on his knee. He twisted the ends and then held it up for my inspection. "Ho ho ho. *Merry* Christmas. Buy some more crap. Come on, line up and buy yourselves a whole lot more crap. Here's something nice. Bought any crap quite like *this* crap recently? Crap crap crap. Ho ho ho. Like that mechanical Santa Claus in the Montgomery Ward's window display. Ho ho ho. Merry Christmas, everybody. Have yourselves all one fucking hell of a *merry* little Christmas, all you poor stupid saps. Line up and get taken, that's what I say, that's what Santa says. We take MasterCard cards and Visa cards. Come on, losers. Line up on this side. Get your money taken on that side." Contrite lay asleep in Rodney's blanketed lap. The car smelled faintly of Contrite's urine, which had a tendency to dribble meekly out whenever he wagged his tail with particular enthusiam.

"Could you do me a favor, Rodney?" I said. I was looking past him in the rearview mirror. Then, at the next streetlight, the police car behind us turned left. "Could you lighten up just a bit? I mean, it is Christmas and all. Just do me a favor and lighten up a little *tiny* little bit, OK?"

MASS

14

Then one afternoon after New Year's Rodney and I were transporting a Panasonic color portable television upstairs to my living room when we discovered Dad sitting very obvious and awkward on the sofa's warped foam rubber. He was wearing a very nice navy-blue Brooks Brothers three-piece suit. He was wearing polished cordovan shoes. He was wearing a white knit tie. "Merry Christmas," Dad said. His golden cuff links gleamed. "Do you remember me? Do you remember who I am?"

Rodney and I very slowly set the Panasonic down upon a pair of wooden crates which were filled with soft drinks looted from the home of some Pepsi executive. I suffered a few moments of light-headed, almost giddy disorientation. Everything about my living room seemed either too large or too small. I didn't know what to say. For a few moments I thought I had staggered stupidly into the wrong home.

"I know I said I'd leave you both alone," Dad said, "but I wanted to bring your presents. It is Christmas, after all."

By way of explanation, Dad gestured at his alligator-skin briefcase with silver clasps. The briefcase was open on the plastic coffee table, revealing festive packages wrapped with bright foil paper, ribbons and blossoming bows. "I even brought a few things for your mom."

"I think I better go," Rodney said. He took the smoldering cigarette out of his mouth and cupped it in his hand. The smoke uncurled secretly through his fingers.

"I'll call you," I said, and then suddenly found myself alone in the house with Dad.

Dad had very distinguished graying blond hair. He had large, ruddy and perfectly manicured hands. His white teeth flashed like spotlights on a movie set. He was a very handsome man, I decided. Even handsome enough for Mom.

We ignored the silence together for a while. The silence inhabited the room like a third presence, or a block of raw marble, implicit with its own hidden Aristotelian form. We smiled at one another. Dad looked at me, then at the Panasonic on the Pepsi crates, still smiling, as if trying to distinguish our relative value.

"That's a nice TV," Dad said after a while.

"Yes, it is," I replied, thinking, This is history. Today I grow up. "It's a Panasonic."

I couldn't just dispatch Dad off without dinner and a few seasonal drinks, especially not after the gifts. A portable CD and cassette player with extendable stereo speakers. Def Leppard. Simply Red. Bryan Adams. Rossini's *Guglielmo Tell* and Strauss's *Der Rosenkavalier*. "Even when you were

little," Dad said, "you always loved opera." There were
dictionaries and desk lamps and an electronic typewriter.
There were shirts and sweaters and underwear and socks.
There was five hundred dollars cash, and two five-hundred-
dollar money orders, one in Mom's name and one in mine.
There was a fully assembled Stingray bicycle, a pair of
walkie-talkies, a crystal radio set, a deluxe Sony Walkman
and two ten-pack boxes of Maxell XLII-S 90-minute blank
cassette tapes. "You can tape directly from the CD onto the
blank tapes," Dad said. "You'd be surprised. It sure sounds
a lot better than those prepackaged tapes you buy at the
record store." For Mom there was a string of white pearls,
and one red rose tidily enveloped by clear cellophane. "I'll
trust you to make sure she gets them," Dad said. "I haven't
any intention of bothering your mother if she doesn't want
to be bothered. I'll be gone bright and early tomorrow
morning. You won't even hear me leave. Here—save these
receipts. If anything doesn't fit or doesn't work or you just
plain don't want it, make sure you take it back and get a
full refund." While Dad spoke he gazed off down the hall
towards Mom's room. He knew exactly where it was, and
that Mom was in it. "I know how she gets. I know there's
no sense trying to make your mom change her mind about
anything."

I fixed almond and broccoli Stroganoff. "It's got broccoli
and carrots and cauliflower in it," I told Dad. "It's got
rice and chick-peas and zucchini. It's got paprika and pota-
toes and seaweed. It's vegan, so if you want a little cheese
on yours, I'll melt some mozzarella in the microwave. I've

115

been getting more and more into vegan food lately, Dad."
I wondered if he was impressed, and passed him the orange
juice. I served the Stroganoff on our house's only two
chipped white plates, allowing Dad the fork while I used
one of our plastic spoons. There was a large hot fire cracking
in the fireplace, and the living room was still littered with
bright crumpled Christmas paper, ribbons, frilly bows and
cardboard packaging.

"Isn't this nice," Dad said. He had taken off his jacket
and unbuttoned three buttons of his vest. He had rolled up
the sleeves of his starched white shirt, displaying the deep
tan of his hairy arms. His teeth flashed, either cavityless or
immaculately capped. "You've turned into quite a respect-
able little cook, Phillip. I used to be a bit of a cook myself.
Back when we were all together, and you were still a baby.
Every evening when I got home from work, I used to cook
meals for you and your mom."

We drank Manhattans beside the fireplace. Once all the
gift wrapping was consumed, Dad walked down to the corner 7-Eleven for a pair of Presto Logs. After my third drink
or so, I loosened up enough to request one of Dad's Marlboros, and even smoked it in his presence. "It's OK to have
a cigarette every now and then," Dad said, examining the
dim ember of his own. "But when you start smoking like
me—two or three packs a day—then you better think hard
about quitting. Otherwise, there's nothing wrong with doing
anything, so long as it's in moderation."

———

116

I couldn't let Dad drive home alone that night. His face was flushed, and there was a remote, insipid smile on his face as he contemplated the fire. I offered him the troughed sofa, the pillow from my bed, and my new sleeping bag. "I'm really glad we did this, sport," Dad said. He had taken off his shirt and loosened the belt of his trousers. He was still holding the empty glass in his hand. "I know it's been a strange situation for a young fellow your age, but I want you to know I'm proud of the way you've handled everything. You're a strong, bright young man, just as I always knew you'd be. Your mom's very lucky to have someone like you who understands and loves her as much as you do. I never expected everything to work out perfect for us, in fact I always sort of expected things wouldn't work out at all. But I'm glad we can spend this sort of time together and just be friends, you know? I'm hoping that whatever happens to all of us, you and I'll always have that."

I don't remember what time we finally went to bed, but my dreams that night were thick with visions of carnival and violence. Strange misshapen men with guns tramped through dry, brushy forests; drunken women danced wildly on tabletops, eventually tearing off all their clothes while crowds of voices roared inarticulately around them like an ocean; alien creatures descended the black night sky in tremendous spaceships, filled with terrible viruses and gigantic, ciliated bacteria which throbbed and pulsed in deep chambers, energizing the ships like fuel. A beautiful white-haired man in flowing white robes emerged from the spaceship and offered me something from one of his soft pink hands. I was on my knees before him. His other hand stroked my brow. Politely, even demurely, I refused; his hand offered again. I refused again, and hard multiple arms grasped me from behind and handed me up to him, bound and helpless on a gold and silver tray. The church below was filled with thousands and thousands of people. At that moment, as I was ritually dismembered before the adoring cries of thousands of dark shapeless figures (the event was being televised, I knew, for modular black cameras weaved the air around me on the platform, attached to long intricate metal cables and winches), I awoke and heard Dad outside in the hallway prying at the lock to Mom's bedroom door. I got up from my bed. I walked to my door and opened it.

"It's very simple, really," Dad said, not even looking up at me in the dark hallway. "You just have to fiddle this little doohickey inside. It's vertical when locked. You've got to fiddle it around until it's horizontal." Dad was sitting cross-legged on the hall floor, wearing only his finely woven and partially unzipped navy-blue trousers and a white Hanes T-shirt. He had an icy, fresh Manhattan in his lap. "Your mom used to do this to me all the time. She was always locking me out when she was upset, but she wanted me to come in and comfort her. It was like a little game we played." He held an untwisted paperclip in one hand, and peered into a tiny round hole of the rattly aluminum doorknob. "These stupid doorknobs are designed to be picked. You used to accidentally lock yourself in rooms all the time when you were little." Dad inserted the paperclip, fiddled around a bit. I couldn't hear another single sound in the entire world. "I'll never forget how scared you used to get. By the time I'd get the door open you'd be hysterical. You'd just be standing there, clenching your tiny red hands, the tears pouring down your little T-shirt."

"Mom," I said, as gently as I could. "Mom, it's Dad. He's coming in."

At that moment the tinny doorknob clicked, and Dad turned it. Dad was still sitting cross-legged on the floor with his Manhattan, and I was standing looking over his shoulder, when the door of Mom's bedroom finally opened.

15

"She's beautiful," Dad said.

"I don't think she's awake."

"Her eyes are open."

"Sometimes her eyes are open but she's not awake."

"Has she been eating properly?"

"I do my best."

"What about vitamin and mineral supplements?"

"I started leaving them out for her, but I don't know if she's been taking them or not."

"What about her cigarettes?"

"I think she's stopped. She's still drinking, though."

"That'll have to stop too. We'll want her eating more fresh fruits and vegetables, more salads. Hot soups and plenty of juices and mineral water. Mainly she just needs a little more exercise, some sun, a few less worries. She'll have to see a good doctor, perhaps a nutritionist."

"I think she's better off staying at home."

"I'll find doctors who make house calls."

"Mom needs her own space, Dad."

"Everybody needs their own space, son. But sometimes their space isn't just their own. Sometimes their personal space infringes on the personal space of other people. That's just a way of saying we all have responsibilities, son. I think I've told you this before, haven't I? We all have very important responsibilities to people other than ourselves." Dad was cautiously approaching Mom like some aborigine trying to console an electric light bulb.

"Dad. I don't think I understand."

Finally, gently, Dad sat on the edge of Mom's bed. The springs emitted a tiny, querulous creak. He wasn't looking at me either. He was looking at the empty bottles of Jack Daniel's and Wild Turkey piled in the corner beside some tattered, outdated issues of *TV Guide*. The neglected vitamins lay heaped on the untidy bureau. Dad's hand gently stroked Mom's stomach underneath the blankets. Mom's face was flushed and serene, like the face of a Madonna. She really was just as beautiful as ever, I thought.

"What I'm trying to say, son, is that very soon you're not going to be alone anymore. You're going to have a little brother, or maybe even a little sister, to take care of. Your mother's going to have a baby."

For a few moments, nothing moved in the entire universe.

"Just look at your mom's smile and you'll know," Dad said. "Just look at her. Your mother always did have a very beautiful smile."

———

121

Understandably, I greeted Dad's arrival with rather mixed emotions, to say the least. Because he was my dad I loved him, but because Mom was his wife that meant he loved her too. There were enough men in Mom's life already, I thought, preparing breakfast for four now instead of two. Dad's lap-top IBM computer was equipped with a modem, which he was usually plugging into the telephone line about the same time I began frying bacon and eggs. Dad sat very erect and deadly serious at the new dining room table he had bought us while strange graphs and data flashed on the amber screen. The lap-top made beeping sounds every once in a while. "What's the dividend yield on that?" Dad asked the phone, when his computer wasn't on it. "Why can't we convert into unit trusts? Sure, sure, but I can't make a living on theories, Harry. I can put credit in the bank. I can invest future expectations. But I can't buy a meal with theories. I can't even buy a good song." Sometimes Dad just sat there and listened. He always wore cleanly laundered and pressed striped shirts and pure wool slacks. His face was always closely shaven. He wore a mild, somewhat suggestive cologne. "All right," he told the phone. He punched buttons on his lap-top, reengaged telephone and modem. "You can start transmitting now." Additional figures and charts emerged on the bright screen. Sometimes Dad paused for a while, watching the cool articulate data emerging from his humming monitor.

Dad communicated all over the world with stock investment analysis coordinators, banking management consultants, corporate holdings portfolio advisers, industrial efficiency maintenance engineers, agents and brokers and realtors and accountants and bankers. While I set our table and prepared our generous meals, while Mom lay in her bed growing my young, immaculate and ethereal sibling in her womb, Dad spoke words not with other people, but with the entire system of language itself. Dad didn't speak things, he spoke systems. Just as systems, I was equally sure, spoke him.

"One of these days we'll have to get you started managing your own money," Dad told me one day, drinking Dos Equis and reading the Dow figures as they unreeled from his ink-jet Epson printer.

"I've got a little money put aside already," I said. I was pouring green soap powder into the soap dispenser of our new Whirlpool dishwasher. I closed the lid, cranked it shut, and engaged the first rinse cycle. "You know, in case of emergencies." The spraying water sounded like static on the radio.

I brought Dad the latest Sears money market statement. I brought him the Dividend Reinvestment figures for my 200 shares of San Diego Gas and Electric. I brought him the joint savings account Rodney and I held at Bank of America. I brought him the savings deposit key. The savings deposit box contained a number of gold and silver coins, some small diversified holdings in various California utility companies, and Mom's diamond wedding band.

Dad examined the various documents for a while. Figures were flashing on the screen of his lap-top, but he wasn't watching them. Every once in a while he whistled a tiny, indefinite melody, not with his lips but with the tip of his tongue against the hard edge of his palate. Ever since I could remember, I always wanted to whistle like that.

When Dad finally looked up at me, his eyes held that reflectionless, distracted expression they usually held whenever he spoke about Mom. He gestured with the various papers in his hand. "This isn't too bad," he said after a while.

"When's he leaving?" Rodney asked. "I thought you said he was leaving yesterday."

Dad was in the shower. I had detached the phone from the lap-top, which had emitted a treacherous little beep.

"That's what he told me."

"Why don't you tell him to leave?"

"I can't just do that. He's my dad."

"Your mom doesn't want him around, does she? What's your mom think about all this?"

"I don't know," I said. "I don't know what Mom thinks about anymore."

In fact, Mom seemed a lot more peaceful and happy now that Dad was back to take care of us again. Because Mom refused to take the lithium Dad's doctor prescribed, every afternoon around four o'clock a private nurse named Syd arrived and gave her an injection. Mom didn't resist with anything but her eyes. "Now be a good girl," Syd said, and officiously posted me from the room. When I returned later, Mom was holding a tiny ball of cotton against the inside of her arm.

124

"It's OK," Mom told me afterwards. Her right arm was growing more and more tracked with tiny scabs. "It didn't hurt at all."

I brought Mom cool pitchers of water, fresh apples and citrus fruits, carrots and hummus sandwiches. Her stomach seemed very taut and smooth when I touched it, like the expanded tube of a tire.

Every morning Dad worked on the dining room table with his lap-top and made phone calls. In the afternoons he took trips into Westwood, Hollywood, Burbank, and often returned long after I was asleep. Mom, meanwhile, lay in her bed, awaited her daily injections and grew more placid and content. This was the only movement left us now, the movement in Mom's stomach. Sometimes she let me place my ear against that smooth and ageless skin. I felt its tiny kick at the world outside. Kick. Kick again. I liked to place my lips against Mom's stomach and speak to it. "When you come out here, don't expect any free ride," I said. "I think I'm gradually working through a lot of the anxieties and insecurities your arrival may bring. By the time you get here, I hope I'll be able to treat you with love and respect for your own individuality. I hope I won't burden you with a lot of silly resentment you won't even understand."

Kick, it said. Kick kick.

"That other noise you hear is just the TV," I said. "Those aren't real gunshots, but just fake gunshots on TV. Sometimes you'll see men fall over dead on TV, but they're not really dead. They're just pretending they're dead. That's the way TV works. People get paid to act like they're people in real life. But they're not, really. They're just actors. On the news, sometimes, you'll see real dead people. I'll have to explain that part to you all over again when you come out. That stuff about dead people on TV always confused me when I was little."

I didn't want any child of Mom's to be confused about anything. Even though Mom had retreated into her formal silence, I wanted our baby to enjoy all the warm immanent attention I had enjoyed, to find itself enveloped in that same constant and imperturbable voice with which Mom had once enveloped me. "In a minute, you'll hear the vacuum cleaner," I told it. "I use the vacuum for cleaning the rug. The suction sucks all the dirt and grime and lint from the floor through this long rubbery hose, and then puts it into this big blue paper sack. Every once in a while, you have to put a new vacuum sack in it. Now that I've finished cleaning, I'm going to be leaving you and Mom for a little while so I can cook us all a good dinner. A pot roast. I roast it in the hot oven. I serve it with vegetables and gravy. That's what you'll be tasting later when it arrives through Mom's umbilical cord. Pot roast and gravy." I never left until after the baby had given me some response. A kick, perhaps, or a hormonal grumble. Then I patted Mom's warm stomach as a sort of telegraphic farewell. I wanted the baby to know I wasn't just talking at it. I was talking with it.

———

I rarely saw Mom and Dad together during the day, but often at night I would awake to the sound of Mom screaming. She wasn't screaming words, she was just screaming. This was another world that emerged in our dark house at night, the screaming of Mom's voice. It was even comforting in a way, formalized by the grammar of our household, by the deep structure of our seemingly eternal and regenerate family. Mom's scream was just a word, like Mom herself, Dad, Dad's lap-top and car, the garage, Mom's TV, and the tiny rough scabs sprinkling Mom's arm like crushed white grains of salt. Covert in my room, I awaited Mom's screams each night like a kiss on the forehead or a glass of warm milk. She'd start screaming for what seemed like hours and hours some nights, as if with her screams she were trying to inflate some enormous circus balloon. One night I went to my bedroom door and peeked across the hall into Mom's room. All the lights were on. The TV was blaring. Mom's face was twisted and angry, flushed and mottled with tears. She gestured wildly with the jeweled compact Dad had brought her from Westwood two days before. She screamed and threw it, and I heard things breaking. She was wearing the white hospital robe Syd made her wear. The robe hung open, revealing her quaking breasts and pale concave chest. And Dad in the doorway too, turning to look at me. He was still wearing his three-piece suit. The buttons of his vest, however, were all undone, and his hair seemed slightly mussed. "Go back to bed, Phillip." Dad's voice was the sternest word. Dad's voice was the world's first word. "Everything's all right. Your mom and I are just talking about something." Then I closed my door again and lay in my cold bed listening to her screaming until Dad finally gave her one of her injections, a sedative prescribed by one of Dad's "specialists," and then presumably they went back to bed together and fell asleep.

16

Suddenly I began to feel different and talk different, as if a different person entirely were developing from my thin body. Dad's arrival really had begun to teach me things about personal responsibility. Every morning I padded softly into Mom's bedroom where I found only the impression of Dad's body in the mattress and pillow beside her. She didn't even look like Mom anymore, but more and more like one of those crepuscular figures who emerged from Mom's dark bed in my most feverish nightmares of Mom. Her mouth hung open, revealing gold crowns and silver cavities. This was Mom's skull, that deep interior world into which Mom could retreat whenever she wanted, and where I could never follow. Mom snored deep in her throat. Her eyes were lashed with milky sleep. "Mom," I said, pushing her shoulder. Her entire body shook with the force of my hand. "Mom, it's me."

The drained hypodermics with their needles snapped off lay in the wicker basket, bedded by bloody Kleenex and balls of cotton. I could hear Dad in the living room, his fingers clacking against the dull cushioned keys of his lap-top. Every once in a while the lap-top beeped or the phone rang. Dad's dark, muffled and sinewy voice would some-times turn underneath Mom's bed like a buried stream or a shifting geological plate.

"Before your dad came I thought everything was starting to work out," Mom said. She slurred a lot. With a Kleenex I wiped bits of saliva from the corners of her mouth. "Before your dad came I was beginning not to be afraid anymore. The spell of my own blood actually made sense to me. Sometimes I even looked forward to having the darkness take me places. It took me down luminous rivers on large rotting rafts and barges. I saw strange birds flying overhead, and the eyes of other creatures emerging from the mucky water. I traveled down the river where twisted houses sat on shores filled with dark men who wouldn't come outside. The dark men were inside whispering about me. They held heavy spears and weapons by their side while their addled women cooked large pots of gristly meat and hung their washing out to dry. The men wore loincloths and streaks of paint on their arms and faces. A few mangy dogs lay around outside the circle of men, contemplating the dim fire. One of the dark men was my father."

Mom sat there and stared into space. She clasped the half-pint bottle of Seagram's against her breast. The bottle was almost empty. Every few nights or so I would go down to our local Liquor Mart, purchase some milk and beef jerky and request cardboard boxes from the bespectacled elderly clerk. While he was out of the store I stuffed bottles of gin and whiskey under my blue Derby jacket. Later, while Dad was away on business, I smuggled them into Mom's room. The bottles, I hoped, would keep Mom company when I couldn't.

"That's not your dreams, Mom," I said. "That's Conrad. That's *Apocalypse Now.* Don't you remember? We saw *Apocalypse Now* at the Sunset Drive-In in San Luis Obispo last summer. At the end, Martin Sheen kills a bull with his ax."

"My father was always a very gentle man with my mother," Mom said. "He had big soft hands. Sometimes he placed one of his hands on my knee when he drove me places in the car. He drove me to doctors mostly. We used to sit together in the waiting rooms and read slick magazines. I always liked *Vogue* best. My father, however, preferred *Popular Mechanics.* Sometimes it seemed like we were sitting in those waiting rooms for days and days, and all the time my father's hand was on my knee. It looked like some sort of animal. I was never really sure if I liked my father's hand there on my knee or not."

"It's OK to be a little confused," Beatrice told me one night on the phone. "It's even all right to be a little afraid. Give your emotions some credit, Phillip. Stop trying to be in *control* of everything all the time."

"I'm not afraid or confused," I said. "I'm not trying to *control* anybody. It's just that I feel there are a lot of decisions being left up to me, and frankly, Beatrice, I don't know if I'm up to them. I love my mom, but I love my dad too."

"You love the idea of your dad, Phillip. You don't love *him*. You love what you want your dad to be. What you want to make him."

"I don't want my dad to be anything, Beatrice. He's just there. I didn't ask him to come."

"Yes you did. You asked him to come."

"No I didn't. He just came, Beatrice. I swear. I didn't have anything to do with it."

"You're afraid it'll all turn out exactly the way you want it to. You never loved your mom, Phillip. You only loved the idea of your dad."

"I love my mom," I said.

"You never loved her, Phillip. You're a man. You're weapons, notions, deeds. You're technology. Your mom's the earth. She's the woods. Your mom's the rain and the wind. Your mom's nature. Your mom's what men's words wreck. Your mom's abundance, but men are cold and hungry. Your mom's life, while men aren't even death, they're just nothing. They're just the cold gray void death presumes to be. Men are the end of space and the beginning of metal."

"I never hear from Rodney anymore," I said. Beatrice's words swirled in my head. I felt dizzy and weightless, like a space-walking astronaut. I leaned against the wall and tried to find assurance in my house's body and mass. I considered tying myself down to something—the kitchen cabinet, or the dining room table. I might float off the face of the planet otherwise, and then I'd never have any idea where I was. A dull ache began to throb in my head. "I call and leave messages, but Ethel says he's out with his friends. She says he'll call me soon. He's just going through a stage, she says. It's just a stage he's going through."

"Rodney's feeling a little hurt, Phillip. I don't think you should have done that to his dog."

I wasn't sure I knew what Beatrice meant. Dogs guarded houses. Sometimes they crawled into your lap and slept there until their owners returned home. My mom had once considered buying me a dog.

"Rodney used to be my best friend," I said after a while. "Rodney and I used to do everything together."

I was only a child. How was I to know what was real and what wasn't? I slept in my clothes—a habit learned during the years of Mom's motion—and often late at night while Mom screamed in the other room I would get out of bed and go to my bedroom door. I had no reason to believe sound was real or even important. While Mom screamed I might also hear the sound of Dad's firm and reasonable voice. "Now, Margaret, you'll wake Phillip." "Especially in your condition." "No you don't. You don't really want to." "I love you, Margaret. I think you know that. Because somewhere in your heart you love me too."

I would sneak down the hall, down the back stairs and into the backyard, where tall weeds towered over me, amber and dead. Morose spiders spun glistening webs in the moonlight, and the high power lines sizzled in the starless sky like

Dad's voice. The power lines were filled with the voices of the world's other dads, calling their sons on the telephone. The world's other dads were real too. They were real people who dealt with real things in a real world. Sometimes they found bits and pieces of the world which were not real, and then they had to make them so, or dispense with them altogether. Things were never as real as they could be for the world's dads. Someday everything in the entire universe would be real, and the world's dads would finally prevail. When that day arrived, civilization and not nature would be rampant. When that day arrived, you could talk to everybody in the universe on the telephone.

"Hello, son."

"Hello, Dad."

"We'll go to the ball game. We'll go to the beach. On the Fourth of July we'll watch the fireworks. Then you and Judy and I'll go to see Judy's parents. You'll like Judy's parents' house. How's your mom? Is there anything you'd like to talk about? Did you finish your science project? Did you get your report card? Is your mom still seeing What's-his-name? Is your mom alone right now? Are you sure? What *is* his name, anyway? I thought I heard a voice. What *is* she doing? What television show? I didn't think that was on tonight. Oh, just in summer, huh? You're sure? You can tell me if she's not alone, son. You know that, don't you? Your mom has other friends now who may even sleep with her from time to time, and that's perfectly natural, OK? There's nothing wrong with that at all. But don't worry. Don't worry one little bit. You don't have to tell me anything if you don't want to. No, of course I don't think that. But just the same. Did she want to talk to me? Are you sure? Do you want to check? Just in case. I'll wait. If she's busy, I'll understand. If she's got company, don't bother her. I'll understand, I really will. I was thinking the weekend after next we'd all drive out to Marineland. Of course Judy'll come too. You like Judy, don't you? I thought so. And she likes you, too. In fact, Judy likes you very much, son. I think sometimes Judy likes you as much as if you were her own son."

I could even hear Mom screaming out here in the overgrown backyard, standing in that strange inverted darkness I found comforting after Dad came to live with us. In Los Angeles the night simmered with its own logic and ceremony. I heard the buzzing earth, the whispering light, the conspiracies of mere matter. Our yard was filled with the ruins of a fallen cement birdbath and weed-sprung brick barbecue. Collapsed trellises, moldering rosebeds, strange, twisted bushes and syllableless insects. I could step into the high weeds and actually feel the language out there, like a human body, like Dad's firm words. Broken cinder blocks, decomposing garbage, the corpse of a sad, tiny sparrow chewed and discarded by some spiteful tomcat. Forgotten civilizations I had read about in books. Mu, Atlantis, Greece, Egypt, Crete, Babylon. Perfect calendars and ritual sacrifice. People torn apart by dark machines. Virgins devoured by sharp blades. The hard inedible fruits of the weeds with hard bright colors. If I owned a telescope and lived on a high mountaintop I could see the stars. Not the stars on a wall map but the stars themselves. Stars exploded and collapsed. They turned and spun. If other people lived in the universe they might be looking at my sun now and contemplating me while I contemplated them. They might be creatures composed of gas or foam or rock or fire. They might live forever. They might love their moms. They might travel across landscapes filled with strange sounds, plants, birds and clouds. They might eat time or fart philosophical propositions. They might live language or speak matter. They may never have heard a single note of music in their entire lives. They might possess the advanced technology required for journeying from sun to sun, but then they might be too lazy and self-involved to bother. Some nights I stood there in the darkness and cried for my lost mom. I was finally beginning to realize that just because I hated Dad didn't mean I didn't love him too. Dad was a house. Mom was just infinite space which Dad's house isolated and defined. Mom was the sadness I couldn't express. She didn't stand a chance. To be perfectly fair, Mom hadn't stood a chance from the very beginning.

17

I began taking less care of myself. I rarely showered or brushed my teeth, and soon grew inured to my own sweet, soury smell. My teeth and gums felt coated with a thin, gritty film. I kept many of the Jack Daniel's bottles I originally pilfered for Mom in my own room now, and dark, amiable boys at the neighborhood bowling alley sold me marijuana, hashish and belladonna. I spent a lot of time alone in my room, listening to Pink Floyd on my headphones. Dylan, Van Morrison, Strauss and Rossini, Handel and Bach. Stoned I felt diffuse and more real. I ate packaged sandwiches, cookies and mints. I watched late-night television. I was growing more solid and real alone in my room while the rich saccharine smoke shaped itself in the air. Men, monsters, sailboats, planets, forests and rivers. Nobody needed me, and I needed nobody either. When I inhaled again from the joint I felt the harsh air filling my abdomen. My blood grew heavy, tranquil and slow, my eyes bloodshot and watery. Sometimes I touched my face, just to see if I was grinning or not. My face felt tight and strained. Every so often I caught myself squinting.

The Jack Daniel's and 7UP always tasted sweet and strong. I could taste it and then, after rolling it on my tongue, swallow and inhale it at the same time. I could feel it going down my throat and esophagus. I could feel it trickling through the twitchy pyloric into the stomach's muscled mouth. The icy drink still felt cool and fresh. Alcohol was pure, like snow. It felt and even tasted like snow, or so I imagined, since snow was one of the many things I had never seen. Sometimes I imagined flaky white snow falling in my stomach. Sometimes I just lay flat on my back on my bed and tilted the icy glass to my lips, leaning it against my doubled chin. Cigarettes tasted better and harsher when you were stoned. I grew filled with a sense of intense well-being. I was not a child, but rather a very wise old man. I had made billions on the stock market, and endowed many large museums and worthwhile institutions. Younger men like Dad were envious of my boats, luxury resorts, gambling casinos, tame striped tigers and insatiable women. My women were of every conceivable nationality and shape. Some of them had enormous breasts, which I fondled one at a time. I made love day in and day out with an impossible assortment of attentive and beautiful women. When I imagined these impossible orgies I placed my hand between my legs. Sometimes it felt hard, but it also felt remote and slightly detached, like a heavy steel pipe or a dictionary. I might start laughing without any reason. I would reach for my Jack Daniel's or my Bud. As I laughed, tears rolled down my face.

"Phillip. Are you all right in there?" Dad's voice roamed outside in the corridor, testing doorknobs and latches, brushing the loose leaves of wallpaper.

"Fine!" I said, and started laughing again.

"What's that smell, Phillip? What's that you're smoking in there?"

The doorknob rattled flimsily. I could hear the tiny lock brace and clack. "Phillip? Why's your door locked?"

"I'm in bed, I'm trying to go to sleep."

Dad's voice waited. Dad's voice was a thing, immobile and immense. Dad's voice lived in the corridor and made lots of money. Sometimes I imagined myself searching through the corridor and uncovering the vast sums of cash Dad's voice had hidden out there. I invested it in Alcoa. My dividends would be nearly eleven percent.

"I don't want you smoking in bed, Phillip. Now put out whatever sort of cigarette you're smoking and go to sleep. I don't want to have to come in there."

"I don't want you to have to come in here either, Dad," I said, and started laughing again. The tears soaked my T-shirt. "I'm putting it out. It's just a cigarette, and I'm putting it out." I was laughing at the impressive mahogany bureau Dad had bought for my room. I had pulled the bureau up and braced it against the flimsy aluminum doorknob. The bureau was exactly like Dad's voice. It was as if the bureau held my door in place on one side, and Dad's voice held it in place on the other.

Dad's voice stayed where it was. It seemed to be trying to confirm something. It was very hard and resolute.

"I love you," Dad said. "And your mother loves you, too. She said kiss you good night."

And then I heard the familiar footsteps, and the door of the master bedroom brushing shut against the shag carpet. I heard the deep breathing house. I heard the distant ticking thermostat. I heard the beetles in the yard, and the electricity hissing in the streets. I heard the stars and the moon. I took another hit off my joint. The tiny pinprick ember flared, seeds popped. A fragment of paper ignited and flashed and its ember drifted up into the air and vanished.

Within minutes, Mom had begun screaming again.

I never wanted to be loved when I was eight years old. I wanted to be crushed by soft massive arms. I wanted to be lifted into some towering embrace. I wanted to be hugged so tight I couldn't breathe. I wanted to be hugged until my eyes watered and my lungs collapsed and my heart popped. I was often awake all night, pacing through the halls and yard of my house, pausing sometimes at the door of my parents' bedroom. After the screaming stopped you couldn't hear anything. My parents' bedroom was perfectly quiet at these times, hollow and hard, as if it had been drained of atmosphere, like some unmanned spacecraft sent off aimlessly into outer space.

"Sometimes I'm not even thinking," I tried to explain to Beatrice one day. "Sometimes I just pace, as if momentum alone compels me. It's like I'm not going anywhere. Just into the living room, the kitchen, down the stairs to the basement, through the icy stone garage, remembering how Mom looks at me sometimes. Her face is flushed and ruddy. She has this insipid smile on her face. Whatever's growing inside her has become wary and suspicious, as if it knows I'm outside waiting. Perhaps it has simply grown stunned by Mom's screaming. 'Everything's going to be OK,' I try to comfort it. 'Once you're outside, you'll have your own crib. We'll put you in my room where it's quiet. You'll eat well. You'll see the sun. You'll reach out your flabby hands and grab my face. You'll wear tiny clothes. We'll hang a bright, intricate mobile over your crib, and it will glitter, so you can watch it at night when everybody else is asleep. You'll stare at the bright mobile and contemplate ideas like motion, light, repetition, difference. These are the best ideas you'll ever have.' "

"Why don't you try taking a Valium?" Beatrice was sitting with me on the garage stairs. The front garage door was hanging wide open. Outside the hot sun flashed across everything: white pavements, white stucco houses, gleaming white windows. Beatrice was twisting the ends of her shirt around one index finger. "If your mom hasn't got any, I can get some from my old man. My old man loves Valium."

"I want things to be different for her," I said. I was staring at the bright sunlight and the wide empty streets beyond my garage door. Inside here the light seemed to radiate from the beamed walls and ceiling, the waxed tarpaper, the cold Rambler. "I want her to be happy. She's a lot smarter than I was at that age. She knows what's out here because I've been telling her. When she's born, she'll find out I've been lying about how much fun it is and she'll hate me."

"Why do you think it'll be a girl?"

"Because I know."

"It'll be a boy, Phillip. Your mother will have about a million sons."

"I want everything figured out before she gets here. I want everything to be perfect for her."

"We all like to think we'll grow up," Beatrice said. "History's the one dream we all try and dream together."

"I don't want to grow up."

"You already have."

"I want to grow down. I want to bury myself in the hard earth. I want to root myself there like a dead tree. I want to entangle myself in the earth's heart so nobody can ever pull me out."

"You'll buy a condo in the Valley. You'll meet a beautiful young woman who drives a silver RX-7. You'll get married. You'll buy a house. You'll have babies."

"No," I said firmly.

"You'll take up gardening, skiing, stamp collecting. Your dad will take you on in the firm. You'll have color televisions in every room of your own house, and video recorders which function by remote control. You'll have second thoughts. You'll wonder what you're missing. Your wife will develop a very distant expression. Her expression will be exploring other continents, even while she's sitting right next to you. She'll start sleeping with other men. You won't even mind that much. Sometimes she'll look very sad. You'll teach your children to be independent, and shower them with presents. You'll tell your children you want them to have all the things you never had."

"You don't know what you're talking about. You don't know me at all."

"You'll buy a sports car, to make yourself feel young again."

"There's things about me I've never told."

"We all never tell things. And we all never tell the same things. They aren't secrets, Phillip. They're conditions. As much as we may all hate to admit it, I'm afraid we all live the same worlds inside."

I awoke every morning with a terrific hangover, parched and aching. Usually I smoked a little grass and took some Tylenol, just to get me started. I watched daytime television in my room. There were game shows which would last forever. They seemed to take up infinite space with their glittering prizes. Each prize bore a cardboard placard describing its retail value. Usually this value was hyperinflated, but always impressive. Sometimes, unconsciously, I tried to add these sums together in my head: $679, $2,807, $99, $3,499. Often the prize was a brand-new car or a world cruise. I imagined myself winning these prizes and taking these cruises. The cruise ships were filled with other boys and girls my own age. There were sundecks and swimming pools, shuffleboard tableaux and billiard tables. We played Ping-Pong and pinball. The girls all wore bikinis. Even though they were only eight or nine years old, they all had very large breasts. I drank Jack Daniel's and Wild Turkey, Stolichnaya and Kamchatka, Southern Comfort and Jim Beam, Gallo and tawny port, Coors and Bud.

"Rodney," I said. I was leaning into my bedroom closet, as far away from Mom and Dad's room as I could get with my telephone receiver. "I need to talk to you. I can't figure this out alone. I need to see you."

"I'm kinda busy."

"Just for a half hour. Maybe I could come over there."

"I said I'm busy."

"But, Rodney—" I was prepared to protest, to cry and shout and hammer and beg. Then I heard Dad's footsteps. They paused outside my door. Then his hand, very faintly, rapped.

"Are you asleep in there? Phillip?"

"I gotta go, Rodney," I said. And then I hung up.

18

"Phillip?"

"What?"

"Are you in there?"

"Of course I'm in here."

"Don't talk back to me, son. I just wanted to see if you were all right."

"I'm all right."

"Do you want anything?"

"I don't want anything."

"Did you fix yourself any dinner?"

"I fixed my dinner."

Dad went silent for a moment. His voice seemed to be gauging things like mass, humidity and weight. "Your mom and I were worried about you."

I refused to dignify such duplicity with a response, and wondered if there was life on other planets. Perhaps it was only microscopic and stupid. Perhaps it wasn't even self-conscious.

"You're spending a lot of time in there alone, Phillip. What do you do in there all day?"

I poured more Jim Beam into my ceramic mug. The ceramic mug said SON, and was part of a family set Dad had purchased at the mall a weekend before. "I read," I said. I reached under my bed for my Marlboro carton. "I'm trying to get some reading done."

"All right, son." Dad stood quietly out there for a moment, like some primitive landmark, or guards outside a condemned prisoner's cell. "I'll leave you alone."

Dad's footsteps beat loudly in the hall. I heard dishes clattering in the kitchen. The television started up in the living room. It was as if my house were inhabited by disembodied sound. Then, after a while, Dad's footsteps returned down the hall. Very softly this time. Then his pocket screwdriver began investigating the lock of my bedroom door. Tick tick. The doorknob rattled slightly. Tick. Tick tick. A flashlight skittered underneath the door a few times.

The locking mechanism on my side of the door was fastened with heavy black electrician's tape. My bureau was braced against it. On the bureau in a small brown bag was the new bolt lock I had purchased at Walgreen's just that morning. I would install it later that night, after Dad was firmly asleep.

The books on my shelves stared down at me like statues or awards, mementos of some former life. They seemed very cold to me now. Books were just the raw matter of education. They were stuff, like coal or minerals. They could be

146

accumulated, quantified and known. I was no longer concerned with the known, but with the process of knowing itself: pure motion, which did not render things known or visible. It did not transport you to any fixed location on a map. It was into the very function of the self that I journeyed now, and like Mom I could only journey there alone. Misery enveloped me with soft black robes. They were warm and clinging. They held me in place so I wouldn't get lost. Misery was my map, my boundary. It held me in place in this world of constant motion.

Whenever Mom started screaming again, I knew that inside she felt warm and safe like me. Sometimes I could hear other voices in there, hidden in that moist miserable world of my own private suffering, the voices of other people Mom and I might have been.

"I'm not a monster," Dad said in my hallway after the lock stopped rattling. "It's not like I don't have feelings too, son. I know I'm the outsider. I know it's going to be hard for you to get used to having me around. I'm not trying to rush things. When I first arrived and everything seemed to be going along so well, I knew this had to happen eventually. You had to start making a lot of heavy psychological adjustments. I just want you to understand I'm doing the best I can, under the circumstances. I'm trying to do what's best for everyone. So I'm not saying you should feel any different. I'm not saying you shouldn't go right on hating me, if that's how you sincerely feel. I would never violate your privacy like that. I'm just saying try to respect my position here as well. Things aren't any easier for me than they are for you or your mother right now." Times like this I thought Dad was a lot like Jerry Lewis on the annual Jerry Lewis Telethon for Muscular Dystrophy. Like Jerry, Dad seemed prepared to go to the most grotesque and inhuman lengths simply to prove his own humanity.

For the first time in my life I was utterly alone. I examined the desultory, overinflated images of naked women in men's magazines. I bought a harmonica which I liked to hold in my hand and imagine myself playing. Sometimes I danced alone in my room, listening to Bruce Springsteen or Joe Cocker on my Sony Walkman. I preferred Jim Beam, but I cultivated a taste for gin as well. I drank and danced until I grew dizzy and surfeited with a thick, swollen stomach, and collapsed on my unsheeted mattress, beating my feet in the air, watching the room swirl around. When it started swirling I knew I might throw up at any moment. That's what the plastic-lined trash bin was for. I lay very still and tried to make the room stop moving. It required an act of intense concentration. It was as if this swirling room was itself a mockery of movement, pulling up through my stomach while the alcohol moved through my blood, lifted into my brain and skull and sinuses. I wanted more to drink and tried to sit up. I knocked over bottles and ashtrays. The gray ashes spilled across my clothes and sheets. There were beer cans everywhere. Everything reeked of gin and cigarettes. The floor of my room looked like the high school parking lot. The world seemed to be growing darker and more desperate. "I don't know where I'm trying to go,

Mom," I whispered, as if she could hear me. "Maybe I'm already there and I don't even know it."

Whenever I was this drunk I couldn't sleep, though that didn't stop me from dreaming. The alcohol seemed to drive my blood and adrenaline as if I were a car, and I experienced strange visions of my progress through the world of light. I proceeded through dense green bushes and foliage. Insects chittered and flashed in the air around me, glancing off my face, biting my arms and neck. I was wearing a large khaki explorer's hat. Native drums beat in the air, and the tortured cries of captive white slave women. I was moving towards a secret road. The road was white and powdery, like the beach of some pristine sea. It extended in two directions. It might take me anywhere. It might take me nowhere at all. The roaring of an airplane filled the blue sky, the beating of helicopters. Everything was blowing around me, even the road's white dust. The plane was descending to take me away, but I didn't want to go.

I would sit up on my bed and reach for the cigarettes, sometimes pulling the glass ashtray or half-empty bottle of Beefeater's onto the floor. I grew obsessed with the idea that dreams were trying to communicate something very real to me, perhaps even the secret advice and admonitions of Mom herself. Dreams were life's base or undergrowth. They were the rich earth from which real life blossomed. The white road, native drums, women crying. The arrival of planes and helicopters, and then the sudden disavowal of everything, the whirling of dust and drums and sky. In the world, perhaps, my progress had been halted indefinitely. But I could still move in my dreams, and in my own feverish and desperate recollections of them. I was learning what Mom had been trying to teach me all along. When I could no longer live in the world, I could live the world alone in my heart.

"I think it's about time you started thinking about somebody other than yourself," Pedro said. "You're not the only person in this world who's lost somebody he loved." Pedro's face was out of focus. I couldn't tell whether it was because of the things Pedro had seen or I had seen that made his face so indefinite like that. "Just think about your dad for a minute. Just think about me, even. Try thinking for one minute about how *I* must feel." Pedro was drifting away. We were all drifting away.

"There's only one thing I've learned in my entire life, Pedro," I told him, though Pedro didn't exist as a body anymore. Pedro existed in the country of my imagination like deep buried formations of rock and basalt. Vast geothermal plates shifted and moved down there. The heat was intense and unremitting. "And I'll tell you, Pedro. I'll tell you what that one thing is." I had spilled Jim Beam down the front of my shirt. I took another long drink from my glass. "I don't care who lost what. I don't care who's gone without love. I don't care about money, or people, or countries, or politics. I only care about me now, Pedro, because now I'm like Mom. Because now I'm never thinking about anybody except myself."

———

Like Mom I journeyed now into my own artificial country—a country much like California, I thought. This country had fields and trees. It was divided into counties, towns
and cities. Several languages had developed over the centuries which, like the landscape itself, were often broken, dis-
continuous and unreal. There were, however, no other
people to be found anywhere. No other people lived and
worked here. No other people dreamed of other countries
like my own. Nothing moved here at all, in fact. Not even
the wind, or the intricately sculpted and often sentient
clouds.

If I were to meet that other woman here, that special
person in my life without a name now, she could not speak
to me. She might not even recognize me. In this privileged
country my imagination had translated everything, even
sensations and appearances, into solid matter, things, atmosphere, nature, earth. That other woman might try to
read to me from books I remembered from my childhood.
Big red and green books with pictures of bears, zebras and
pelicans in them. That other woman might try to speak to
me, feed me, smell me, touch me, but she couldn't do anything about it. She had lost all pretensions to form and
become only a force, a dream of falling. She could only run
through her accustomed routine—combing and petting my
hair, stroking my warm stomach, clipping my toenails—
while I drank my gin and tonics and contemplated the walls
of my country. My country had fissures and cracks, exposed
plaster and wiring. The doorknobs didn't always correctly

articulate with their locks. The windows were often painted shut, their curtains torn and dingy. Dishes were chipped, and glasses were filled with spongy cobwebs. None of the silverware matched. If you stood on a chair and gazed into the highest cabinets of my country you could just see, in its very furthest, dingiest corner, something tiny which didn't move. It was gray and shapeless. It possessed no smell or texture. If you looked at it too long, it wasn't even there at all.

I no longer wanted to live forever. I wanted only to burn intensely in the sun and extinguish, seedless and unforlorn like some unpopped pinecone. I began taking foolhardy walks at night into the worst parts of town, where strange dark men and women stood in doorways which were often plastered over by promotions for rock concerts and record albums. If I possessed no home, then logically my only home lay everywhere. The streets were filled with refuse. Garbage cans were overturned, and the wild dogs did not approach you. The dogs always looked at you over their shoulder as they paced anxiously away. They seemed to be calling you. Perhaps they were just hoping you would call out to them. They suffered terrible skin rashes and limped, like many of the strange men and women who stood loitering in doorways, or pushed large shopping carts about. These parts of the city were like some postnuclear landscape. These broken people had survived the extinction of an entire civilization. There was something admirable about them, about their rashes and strange growths and misshapen features.

They drank very cheap wine out of paper bags. Sometimes I stood and watched, and they offered me some. I always refused. It was not because I was afraid of contamination. It was because I was afraid I might contaminate them.

"How you doing, Johnny? How you doing, baby?" The sexless old woman had a faint gray beard and a pale face. "You coming home tonight? I'll fix your bed. You come home tonight and I'll fix your bed." Her fingernails were dirty and untrimmed. She held in her hands a gruesomely stained and tattered paperback copy of *The Hite Report on Female Sexuality.*

"Johnny's dead," I told her. "Johnny died last night in the hospital of tuberculosis. He asked why you weren't there. He asked why you weren't even there to say good-bye."

Sometimes I would walk all night, through unmapped and remote parts of the city which might have been only dreams. I remember long dark alleys filled with yowling cats. I remember men dressed as women, and women dressed as men. I remember bodies asleep or dead, and when I touched them with my foot they didn't stir or respond in any way. Every few blocks or so I might find a well-lit liquor store open where I could purchase cigarettes from an indifferent clerk who watched X-rated movies on his videotape machine. I remember horrible-looking men who called out to me when I passed. Sometimes they might start to follow. Their bodies seemed to have collapsed inside their matted gray clothes. Sometimes they emitted terrible, half-human sounds, and I would run away. I don't remember what was real and what wasn't real during those long restless walks I took far away from my home. I only remember myself moving deeper into the buried countries of my imagination, where one found one's way purely by instinct. I could never be sure where I might arrive next. By now, of course, I had said good-bye to California altogether.

"It's not a plot or anything, Phillip," Beatrice told me. "You're not history. You're not what things happen to. You're just a little kid, Phillip, who's got a number of severe personal problems right now. I've never suggested this sort of thing before, but maybe you should see a counselor. Perhaps you should seek professional help."

"Fuck you," I said. This was the rage I loved. I could drink and rage like this for hours, if only someone was there to help me. "Fuck you, Beatrice. I know what I'm fucking doing. I don't need any fucking body. I don't need you or Rodney. I don't need *any* fucking body to tell *me* what I should be doing."

I wasn't even listening for the click of the extension. Silence emerged suddenly from the telephone line like an official sanction. It was the world of electricity. It was the world of pure force.

"Fuck you, Beatrice," I said. "Just fuck, fuck you."

Eventually of course even Beatrice stopped returning my calls, and at night I nervously explored my neighborhood and the dark, abandoned playground of the local elementary school I had never attended. On this grass, and among these swings and monkey bars, children my own age played every day at appointed times. They ate their lunches on these splintering wooden benches, underneath these deliberately shady (and somewhat smog-tarnished) elms and sycamores. They took their schoolbooks home and, after they had completed their homework, they were permitted to watch TV, or invite friends over. In class they constructed synonym wheels with colored paper, scissors, paste and a dictionary. They banged wooden blocks together. They presented staged dramas about ecology, history and tooth decay in the echoing, cathedral-like auditorium. Tanbark and blacktop, hopscotch and four-square diagrams, softball diamonds and backstops, volleyball chains whispering against tarnished steel poles. At night, encircled by the sloped hills and hedges and the higher streetlamps, the shadows were gigantic here. For me childhood seemed like a sort of ghost town. It was ancient history. Its remnants hung about like some forlorn and noble calculus, Stonehenge or the pyramids.

It was as if I didn't even exist anymore. I didn't have a home. I didn't have friends. I didn't even have a mom. I just had the shadow of him, him in my house, him with the key to my car, him with the checking account now, and my mutual funds, and my T-bill account, and my government bonds and silver certificates. The emerging shadow of him with the MG, the beeping calendar watch, the Filofax, the Ralph Lauren cologne.

It was hot out here, even at night. The smog and city lights of the San Fernando Valley extended into the sky, absorbing stars, galaxies, even notions about the way worlds worked. It didn't even feel like loss anymore. It didn't feel like dispossession or grief.

"I can't go back there," I said out loud. "But I can't go anywhere else."

On the suburban streets and avenues surrounding the schoolground car doors slammed shut. Entire families were coming home together from movies, pizza parlors, bowling alleys. The doors of houses and garages opened and closed as well. Lights went on and off. A dog began to bark.

"Hush, Luke!" someone demanded.

"You can mope and feel sorry for yourself all you want," Pedro said, obscured by the bristling darkness. "But I don't think that's going to change anything, do you? I think it's time you started taking responsibility for yourself, and stopped blaming everything that happens to you on everybody else."

"I don't hate him," I said.

"But you want him out of the way."

"I understand how he feels. I know he just wants to help."

"But look what he's done to your mom."

"He transferred most of my stocks into money market accounts at just the right time. He saved me almost a thousand dollars."

"You can't even control your drinking anymore. You can't control when you drink, or how much."

"I'm just like him. Beatrice is right. Dad and I are exactly the same man. I'm like a homunculus, Pedro. It's like Dad's the body, and I'm the DNA. That's the unit measurement of life. That's where all the most complex, unconscious decisions are made. They're made every day when we're not looking down inside the DNA."

"He's taken the keys to your car. He's locked your car in the garage."

"He's a very successful businessman."

"He won't let you touch your own money."

All of the sky's stars were invisible tonight. The sky was only a haze of city lights. And this bristling noise—the noise of high crisscrossing power lines and tall streetlights.

"I don't know what to do, Pedro."

"You know what to do."

"I'm completely confused. I can't think straight anymore."

"You know what to do."

I turned. The schoolground seemed to be glowing. It wasn't like night so much as like night on some high-tech movie set. Invisible machines operated everywhere. Hidden technicians monitored, taped, replayed and edited. Truth could be collapsed and disarranged. Life was not fact, but montage. I might even be an actor playing somebody else's role. My mind might be the stage upon which some cultural drama played.

"You tell me, Pedro." I couldn't see him anywhere. "You tell me what to do."

The entire universe took a long deep breath. I was part of that universe. These planets, these stars. This was my real home.

"Kill him," Pedro said. "Kill him. Kill him tonight in your mother's bed, just like Hamlet. Kill him at breakfast— I'll show you how. You can run him down in your car. You can poison or cut him. You can strangle him in his sleep. But kill him. Kill him anyway you can. Kill him now, Phillip. Kill him now. Kill him now."

CHEMISTRY

19

"Good morning, Phillip."

Dad was wearing a three-piece pinstriped navy-blue Brooks Brothers suit. He was wearing leather Rockport shoes, a white knit tie, matching 24-carat monogrammed cuff links and tie pin. Everything looked really good on him. My father, I had to admit yet again, was a very handsome man. He looked much younger, in fact, than I felt. He took up his folded cloth napkin and sat down. Each morning I would begin setting the breakfast table as soon as I heard Dad's shower start up in the master bathroom.

"Decided to join us back in the real world again, huh, son?" Dad poured ringing Cheerios into his blue ceramic bowl. "I'm glad."

Cheerios are a happy cereal, I thought. I hadn't slept all night, and was sipping my fourth cup of black coffee. Cheery Ohs. Cheerios.

Dad opened the massive *Times* and deftly disengaged the Business section. I was gazing dully at the muted yellow standing lamp in the dining room. My mind was keen with adrenaline, but my body sagged.

"Gold's up," Dad said behind his paper.

"I've been thinking, Dad. I've been thinking a lot about us lately. You know. You, me, Mom."

Dad's paper minutely rustled. The rest of the *Times* lay massive on the table like a fish tank. The front page said things about the Middle East, unemployment, presidents, Taiwan. Someone had survived an attempted assassination. Someone else had just been born.

"I'm not proud of the way I've been behaving," I said. I was still staring at the yellow lamp. My body seemed to be shifting. Or perhaps it was the house that was shifting under me. "It's not like I'm trying to be selfish. I know things haven't been easy for you either, Dad. What with Mom pregnant and all, and me just sitting in my room feeling sorry for myself. I haven't been helping much around the house. I haven't been keeping up the yard." My first and only substantial memory of Dad was much like this. We had been divided from one another by the thin fabric of Mom's stomach. He had said things to me in his deep voice. The words had reverberated in my warm, amniotic placenta like chimes in an iron bell. "I'm going to try and make it up to you," I said. Now my words reverberated, now my words filled things. The Formica table, the plastic chairs, the dishes in the cupboard, the bills on the microwave. "I think it might make things better for all of us."

With a flourish, Dad shook the leaves of his newspaper back into conformity and refolded them. Dad's face was flushed and slightly sunburnt. I was beginning to suspect a sunlamp in Dad's office. He was smiling. He, too, thought things would soon get better for all of us. He suggested a movie that afternoon; he would take off early from work. Or a baseball game. The Los Angeles Dodgers, as I understood, were our home team.

Yes, handsome. Very youthful.

I told him I'd think about it—I had many errands to run that day. Maybe tomorrow—or better yet, next week. Sometime that afternoon there were a few things I needed to pick up at the hardware store. Dad loaned me his Visa Gold card.

We were both still smiling very energetically at one another when Dad left for work.

The body, I have often thought, is like a promise. You keep things in it. Those things are covert, immediate, yours. There is something lustrous about them. They emit energy, like radium or appliances. They can be replaced, repaired or simply discarded. The promise of the body is very firm and intact. It's the only promise we can count on, and we can't really count on it very much.

"We'll need this," Pedro said. We were touring the hard aisles of Ace Hardware and Plumbing Supplies. I was pushing the Ace Hardware shopping cart, on which one eccentric wheel wobbled and spun contradictorily. "We'll need this, and this, and this and this. We'll need one of these. And we'll need one of these." Finally we decided on a large steel toolbox with trays which extruded when you opened

163

the toolbox lid. I was very impressed with Pedro's knowledge of tools. He knew which had the sharpest blades and
points. He knew which ones were built to last.

"My, my," the checkout lady said. The checkout lady
had very gray hair. She wore an official red Ace Hardware
apron and name tag. The name tag said her name was Doris.
"You look like you're going to start your own little business." She held herself with such ardent restraint I knew
she could barely prevent herself from patting my "tousled"
head. She took Dad's Visa Gold card.

"Did your father tell you you could use this?" Doris
asked, still beaming her flawed white dentures.

I offered to show her my ID.

"No, no. That's fine." Doris ran the card through the
gutter of an electric scanner. "You're a very industrious
young boy," she said. "How old are you, dear?"

"I was eight years old last November," I said. Then I
carried the toolbox back up the street to my house.

The tools fitted snugly into their appropriate compartments. The lid of the toolbox closed securely. It was like a
body with its tidy organs hidden inside by warm, glistening
envelopes of tissue.

"Where'd you go, Pedro?" I asked.

"You didn't need me around. You had your dad's credit
card. You had all you needed. You didn't need me."

Only after I urged him repeatedly did Pedro show me
how to fit the gleaming hacksaw blade into the firm steel
handle. Outside it was very hot and sunny. Outside, the
sun-dazed birds in the trees didn't make a sound. Later I

stored the pregnant toolbox in my closet under a heap of
new clothes Dad had been buying me over recent months.
When I closed the door, I could still detect the gravity of
it there. It seemed charged with its own imminence. It
cooked there, like the dashboard lighter in the days of Mom's
luminous motion.

"If Dad's body was a house, Mom, what kind of house
would it be?"

"It would be a very big, safe, well-built house with stone
walls and turreted balconies. It would be made from the
best materials. It would be perfectly designed to benefit the
needs of its specific occupants."

"Would it have heavy doors?"

"The heaviest."

"Would there be alarms? What sort of alarms, Mom?
Sonic or wire? Would they be hooked up to some private
agency, or to one of those shrill fire bells? Would there be
dogs, Mom? Would private guards check the place out
whenever people were away?"

"Sonic," Mom said. "Everything in this house will be
very clean and compact. There won't be any wires or cords
to entangle or confuse you. Everything will be in its proper
place."

"Will I have my own room?"

"We'll all have our own rooms. Our own bathrooms and
baths. We'll have private brick fireplaces and balconies. We'll
have well-equipped bars and portable color televisions in
every room hooked up to satellite-dish reception antennas."

"Will there be a yard?"

"A vast green expanse of it. You won't even be able to see let alone comprehend its gate. There'll just be a distant green vanishing point. You'll think you're losing your sight. You'll feel like you're living on your own little planet, a planet with abrupt horizons and green earth. Your eyes will strain and water if you try to take in too much of it at once."

"Is it in the country?"

"It's in the city and the country. That's how big it is."

"At night can you see the stars?"

"You can see the white buzzing band of the Milky Way. Every star in the sky shining at once. It won't seem like light, but matter. It will weigh against you like mass. It will shine like gold. You'll be lying on the green grass at night and you'll try to reach out for it. You'll try to touch it. Your hand will ache, just thinking about it."

"I can enter Dad's house any time I want."

"Of course you can. It'll be your house too. You'll have your own key."

"I'll have keys to all the rooms."

"Except the most secret ones."

"But even then, I can peer in the windows. I'll be allowed in the secret rooms when I'm older."

"When the time comes, you can discuss that with your father."

"I only want to go into rooms Dad's already been in himself."

"You'll have to be careful."

166

"I'll find thick complex networks of lymph and artery and tissue there. Fatty deposits, moist and cellular, like the eggs of fish or amphibians. The hard moist marrow filled with yellow matter. Renal ducts and spongy scoops of liver. The hard muscled heart. The body's stringy muscles knitting ribs and shoulders and stomach. Bones articulated with tendons, cartilage, gristle. Bones articulated with other bones."

"That's something that's all none of my business," Mom said. She was staring at the television screen. Neon cash amounts flashed on an enormous multicolored board. The faces of celebrities beamed and flushed. One hugely happy and obese female contestant began to cry, surrounded by gigantic photographs of all the wonderful, exotic places she would soon be visiting. "That's all something you'll have to discuss with your father."

I felt the cold strobing black atoms humming around me in the darkness as I swayed back into my room through the swirling hallway. It was just Mom and me again. Mom in her room with the baby growing inside her like a secret, like the secret promise of bodies, and me in my room across the hall with Pedro, spinning our thin schemes of savagery and mutilation, trying to push Dad into the future again where he belonged, into the future's deep dark earth where the black atoms hummed and spun like elementary planets. Mom and I had grown so far apart we could understand one another again. Ours was a cellular complicity. Even when we lived in different countries and spoke untranslatable languages we still knew each other. Mom knew what

I had to do, and approved with a covert dense resignation. It was an approval that made Mom's body hard and perfect and safe.

"That's a Conready steel file," Pedro said, showing me how to hold it. "Feel that weight. There's a lifetime guarantee on that sucker. You want to know how many steel tool manufacturers offer a lifetime guarantee on their merchandise? I'll tell you. Not too damn many, that's how many. Not too damn many at all."

I was suffering giddy delusions of grandeur. I considered writing the local newspapers and confessing to them all my plans. Then later I would send them photographs of what I had done to Dad while they stood helplessly by, pretending their authority over world events. I felt like an astronaut who had just returned from deep space and the exploration of rich worlds inhabited by aboriginal cultures and convoluted blue cities into which random asteroids often crashed. My mind and knowledge were astrally privileged; I had returned to live in a world of tiny, ineffectual ants. Just as the baby was assembling itself in Mom, I now felt Mom assembling herself in me. Everywhere I went I went with Mom's implicit consent. Everything I did I did at Mom's unspoken command. I felt incalculably brave. I felt invulnerably correct. I felt like science or politics. I had broken the cipher of eternal language; I was learning new words about the real universe, and with this eternal language I would live forever. Words were what mattered, not bodies, not things breathing, vulnerable and vile. Life was a hard word, or a sentence filled with hard words. The process of

168

living was not a problem of biology but of grammar. One simply needed to know how to arrange oneself within the proper sentence. One simply needed to comprehend not one's substance or actuality, but rather one's relationship with the world's other hard objects. One night I was filled with such weightless soaring arrogance I even called Beatrice and confessed everything.

"You can't fool me, Phillip. Don't call me up at home just to hand me this line of nonsense." I heard the pronounced puff of Beatrice's lips against her filtered cigarette. She exhaled the smoke with a long theatrical sigh. "You love your dad. You'd never do anything to hurt him, and you know it."

"I hate him." I was finishing the last of Dad's Wild Turkey, garnished with a splash of crushed ice. "It's all going to work just perfect. I know exactly what I'm doing, because I've done it all before."

"You *can't* hate him, Phillip. He's your *dad.*"

"I've got it all worked out. I know exactly what I'm going to do."

"You're going to kill him," Pedro whispered, somewhere in the dark.

"Isn't that just like a man," Beatrice said. "To say he knows what he's going to do when he hasn't got any idea what he wants or even who he is to begin with. Who the hell are you, Phillip? That's the question I'd like answered. Not what stupid things you're going to *do.*"

"Kill him," Pedro said. "Kill him, kill him, kill him."

"I hate him," I said.

169

"*You* hate *him*, Phillip? That's just what I mean, isn't it? Man's myth of intentionality. *I* do things to *you*. Predication. Subject and object. The dream of perfect cosmic grammar. Well, dream on, kid. Dream on till you're old and gray. Because you're old already, Phillip. You're already older than your own dad. If you want to know the truth, I think I like your dad better. At least he *tries* to make things work without bullying everyone all the time, like you or Rodney. In fact, Phillip, I think I like your dad a whole lot more than I'm beginning to like you."

The black atoms swarmed and rushed around me again and again. They were my private, internal atmosphere. They would take me back into the darkness with them again very soon. "Fine," I said. "I don't really care what you think, anyhow."

"Let me put it to you this way, Phillip. *You* don't matter. Neither does your dad. We're all nothing but heat, motion, gravity, sound, history, light—that's all that matters, Phillip. Just force and stress, time and matter. Start thinking about those things. Get outside your feeble, crowded little brain for once in your life and try looking at the big picture, will you? You guys all like to think you're such hotshots, you're all in such control of everything. Well, you're not. You're nothing but a bunch of dicks, that's all you are. Us women, on the other hand, we're what you call heterogeneous. That means we're everywhere, everybody at once. We're both good and bad, right and wrong. We're the great resolvers of conflict, Phillip. We're like octopuses—because we'll swallow *anything*. Even men. Even battling and forlorn men like

170

you and your dad. You guys try so hard to be subjects, characters, things, you forget us women are the whole story. We embrace you all. What you really want to destroy is women, that story of yourself you can't control. Women are what you really hate, Phillip, not that poor dumb jerk of a dad you've got. I've been reading a lot lately, Phillip, since we broke up. French feminists, existential Marxists. I'm teaching myself French so I can read Sartre's *Critique of Dialectical Reason*—much of which has been improperly translated, from what I understand. You should learn to speak French too, Phillip. Then we could talk French to each other over the phone."

"I'm sorry I called," I said. I could hear something clacking wetly in Beatrice's always lugubrious mouth. I could even hear its faint reverberation against Beatrice's crooked teeth. Hard candy, I thought. Perhaps a Tootsie Pop.

"Do you miss me?" Beatrice asked after a while. "Have you missed me since I haven't been around?"

There was more to the universe than light, gravity, mass, history, motion and sound. That's where she was wrong. I was in the universe, too. Me and Pedro.

"Sometimes, I guess," I said. "I guess sometimes I miss you a little bit."

We would have to move quickly—finally Pedro and I agreed on one thing. Dad had begun talking about returning Mom and me to our true home.

"As you know," Dad said to me one night over dinner, "I haven't wanted to rush things up till now. I didn't see any hurry. But now I don't see any need to waste any more time around here, either." Dad indicated the thin dismal living room, made even more sad and depleted by the bright new furniture and drapes Dad had installed since his arrival. "I don't think I'm just speaking for myself here, but let's face it. It's pretty depressing, wouldn't you say? It seems to be getting on everybody's nerves."

"It's just a house," Mom said. "It was just the first thing I could find when I needed one."

Mom was seated across the table from Dad. She was wearing a broad-waisted cotton summer dress. She gazed emptily at the curtained window behind Dad as she chewed her Chinese noodles. Mom had only lately, at Dad's polite insistence, begun taking meals with us.

"Our real home's still waiting for us," Dad said. "There'll be a nursery for the baby. There'll be room for a live-in nurse to help your mom out. And you, sport. You."

Dad offered me the rest of the cashew chicken, which I perfunctorily declined. I had suddenly lost my appetite.

Dad forked the remainder onto his own plate. "You'll be going back to school. You'll have a nice room, and a proper library to study in. You know I bought the Britannica School Edition for you since you left? I put it up in your room already, along with a few other things. A word processor, your own video machine, some classic movies on cassette. I think you should start filling in some of the gaps in your education—I'm talking popular culture, here. The entertainment industry. That's the business I've always expected might attract you someday. Film, television. Cable's opening up a lot of new ideas in marketing. The production end, that's what you'd be good at. Technical equipment, where the real money is." Dad poured himself the rest of the lapsang sou-chong. He gazed into the tepid brown water, as if he were reading the arrangement of leaves at the bottom of his cup. I wondered if they said anything interesting. "I think it's time we got on with our lives," he concluded. "I think it's high time we all stopped messing around."

2 0

"Hi, Ethel. How you doing?"

"Phillip. Phillip. Oh, isn't this nice. Isn't this nice to see you. *Phillip.*"

Ethel's voice sounded and recoiled at the same time, stepping her lightly backward into her immaculate living room. Even as she called my name she seemed to evade me. I stood on the sunny porch and she in the shadowy doorway. The contrast made her appear at once firm and unfocused, as if every particle of her was being diffused by some foggy outdoor film screen. She looked like an apparition, like one of my recent dreams of Ethel.

"Is Rodney home?" I said. "Do you think I could see him?"

As she was letting me into the hall Rodney's voice abrupted warily from the top of the stairs. "Who is it? Ethel? Is it someone for me or what?"

"I think it's good you've come over to make up," Ethel whispered. Quickly she began handing me whatever was

174

available on the dining room table. A plate of crescent-sliced, crustless sandwiches. A bag of Cheetos. A couple of cold Buds. "It doesn't matter who starts an argument, now, does it?" she said, winking. "It just matters who's man enough to make up. If you're friends with somebody, then sometimes you've just got to swallow a little of your pride, don't you, Phillip? It's good to have you back, dear. I think he's missed you. I think maybe we've both missed you."

I really wasn't up for Ethel right now. Every gesture she made suggested she and I shared some secret agreement which either excluded or diminished Rodney. I wanted to tell her that she didn't matter. That the friendship Rodney and I shared not only superseded her, but was actually none of her business.

I carried all the stuff up to Rodney's room, balancing it against my stomach. I knocked lightly with my knee.

"Conviction's all we lack," I said. "That determination not simply to be ourselves, but to be *anybody*. We should carry our conviction like a hammer. It doesn't matter what we build. It only matters that we *act*. It only matters *that* we build." I was soaring now. I was thipping across the lane dividers in Mom's luminous car again. I saw Buellton. I saw Fresno. I saw Salinas. I felt Mom's voice rising in me strong and intrepid for the first time in months. I would never die. Mom would never leave me. "We're like armies of men, political nations, the corporate arrangements of cells, tissues and bodies. We're not children, Rodney. We're the world. We're greater than the world, because we can make it be anything we want it to be—no matter who tells us otherwise. We're all that matters, Rodney. All that matters are our strategic situations, and the tactical *stuff* we use in order to get where we want to go, in order to take what we want to have. Where we are, what we get, how we get it—*that's* all that really matters. We act together, Rodney, just like always. You and I. It's not like we have any choice. It doesn't matter if you like it, or if I like it, or even if we like each other. It just *is*, Rodney. You and I just *are*. We're stuck with each other. We're friends for life."

Rodney had hardly touched his tidy fragment of sandwich. He examined it distantly now where it rested on his jiggling knee like a trained hamster.

"I liked that damn dog, Phillip. You knew that, too. It wasn't like you had any right. It wasn't like it was your dog or anything."

I could hardly recognize Rodney. A blue pentangle had been tattooed around the frame of his left eye. His hair had been shaved back to reveal a high, shiny forehead; it was tonsured and dyed a streaky, phosphorescent green. His room was filled with books on voodoo, black arts and magic. A cone of sandalwood incense burned on a tiny brass devotional tableau decorated with the bodies of naked, writhing women with serpentine hair and pointed breasts. From his left pierced ear dangled a silver earring intricately carved with the skull of a leering baboon. The floors and furniture were littered with various lurid paperbacks with bright red colors depicting flames, apocalypse, demons and witches and complicated demonic symbols. The books were entitled things like *Hell Town U.S.A.*, *Cry the Children*, *The Book of*

Satanic Myth and Lore, UFO Sightings Unveiled, and *Sydney Omar's Guide to Astrology: Taurus.* "I'm glad you've finally developed an interest in reading," I said after a while. "It always helps to find a subject that interests you."

"I guess I've just been bored," Rodney admitted later, while I was searching through his closet and rearranging the piles of moldering laundry I found there. "Bored bored bored. Jesus Rice Krispies I've been *fucking* bored. Getting up, eating Ethel's lousy goddamn breakfast, going to school. And the teacher's droning on about this and that and the other thing. I try to tell the teacher, you know. I don't give a fuck about geometry or English. Like I'm probably going to drive a truck or something when I get out of school. Join the army or something simple. I'm sure in the *army* they're all going to be wondering what an acute angle is. I'm sure I'll make lots of friends driving my truck because I can diagram some lousy goddamn sentence. And then after school I'm free, right? What's that mean? I go down to the bowling alley or the shopping mall with my friends. We scope the girls, smoke a little doobidge, maybe a tab of acid every now and then. But that's not really living, is it? I mean, if that's living, then excuse me right now. I'll go out and put a bullet in the old brainpan. But if that's *not* all there is, right, well, maybe there's something I could do a little less radical, like, you know. I don't mind life or anything—I'm perfectly willing to give it a try. So what the hell, I figured. I'm sick of school, drugs, this goddamn oppressive house of Ethel's and all. Maybe it's time I experimented a little more with my life, took a few more chances. So that's when I decided to become a warlock. To master the satanic arts of black magic. Devil worshiping, for you laymen. I want to learn to master what they call the black arts."

I moved aside cardboard boxes filled with Marvel and DC Comics, dismembered football and hockey uniforms, baseballs and baseball gloves. Then, behind some crumpled *Playboy* magazines, I found it.

"It's all relative," Rodney said. "Black magic's no 'better' or 'worse' than white magic. It's not like one's 'good' and the other's 'evil.' It all just depends on what side you're rooting for. In other words, it's all relative. Black magic can go places white magic can't, that's all. Satan's not any more evil or good than God, he's just trying to move in on God's territory, like General Motors or Chanel. Everyone faces competition—that's what makes the strong stronger. That's why civilization gets better and better instead of just falling apart. I say use what you've got in this world, because nobody else is going to give you anything *they've* got. Use what you've got, or else the other fucker's going to use what *he's* got on *you,* and I'm not kidding. I think you hear me, Phillip. I think you know where I'm coming from."

A Judas Priest album was playing full blast on Rodney's stereo. I couldn't make out the lyrics very clearly. It seemed as if they were screaming, *Retribution, retribution, retribution, retribution . . .*

"Could you turn that down?" I asked, unlatching the slightly oxidized steel clamps and unfolding the chemistry set atop Rodney's rolltop desk. I pushed some of Rodney's

178

soiled paperback books out of my way. Aleister Crowley's *The Book of the Law*. John Knox's *Satan's Women: A Guide to the Pentagram*. A couple of James Herbert and Stephen King novels. L. Ron Hubbard's *Dianetics: The Modern Science of Mental Health*.

"I've even managed to explain this all to Ethel so even *she* understands where I'm coming from," Rodney said. "And Ethel, as we all know, is a stupid cunt."

"Now," I said, "how about a little more light?"

21

I couldn't just go and kill Dad, say with a gun or a knife or a bomb. He couldn't be obliterated, like propositions or houses. He was far too vast and remote to be assailed by small hands and arms such as mine. No, if I wanted Dad out of the way, then I had to deploy Dad's strength against himself. I could not conquer, Dad could only succumb. I could not be the agent of Dad's death, only its engineer.

While I tapped a few intricate crystals of sodium into a beaker, I suggested Rodney turn on his cassette player.

I would have to insinuate the diffuse, inorganic world of chemistry into Dad's body while Dad wasn't looking. Rodney put on the Grateful Dead, and I began furiously assembling compounds with which, over succeeding days and weeks, I regularly began dosing Dad's coffee, cookies, roasts and steaks. Dad always said he and I should go more places together, so I proposed we go everywhere, everywhere at once. We were going to journey into the real scheme of life, Dad and I, into life's basic and molecular stuff. The assemblies of atoms and molecules, that systematic world of electrons which orbited and contextualized mere physics. Appearances, behaviors, properties, symbols and formulae, enumeration and analysis, polymers, fuels, oxides and energy. I was going to return Dad to that world where he truly belonged, that fundamental world of basic particles which breathed underneath our realer world of mere events. Meanwhile, in Rodney's room, the stereo played:

Many rivers to cross,
And I can't seem to find
My way over.

Soon Dad was suffering colonic spasms, flatulence, rashes, dizziness, occasional vomiting, boils, sore throats, hemorrhoids, blurred vision and acne. "I don't know, sport," Dad said, one hand resting covertly on his stomach. Mom was still patiently chewing her prawn salad. "I don't know if I feel like dessert or not." I had been experimenting with sodium compounds that night. Sodium sulphate in Dad's mushroom soup. Sodium thiosulphate in Dad's tandoori shrimp.

"Don't worry about it, Dad," I said. "I'll do the washing up. You get yourself some rest."

Mom, impressively large now, sat resolutely at her place, eating everything in sight. Once the salad and entrée had vanished, Mom commenced tearing into the sourdough french bread and margarine.

"I think something's bothering your father," Mom said,

after Dad had gone into the living room to lie down. Mom was crunching french bread in her mouth, scraping crumbs from the corners of her mouth with one long fingernail. She was staring off into her private country where our baby was lifting itself onto its hind legs and uttering its first hesitant vocables. "Your father hasn't been sleeping well," Mom said. "Sometimes he wakes the baby. Sometimes he's so restless I can't sleep, I can't even relax." Mom made soothing motions against her stomach. "I may begin asking him to sleep on the living room sofa." Mom was wiping the doughy center of the bread at her plate until the plate was white and dry like a bone. Then she put the soft, moistened bread into her mouth, a patient, animal expression on her face, complacent but alert.

"I guess you can say I started taking a serious interest in Satan about six months ago," Rodney said, while I heated random substances in a beaker over a thinly glowing can of Sterno. "But that doesn't mean Satan hadn't been in my thoughts long before, or that I wasn't in some important way already under his influence even when I was very small, even before I was old enough to talk or read. I think I always knew about Satan, but it was only unconscious knowledge, if you know what I mean. There's lots of knowledge that's important in this world, and you don't necessarily have to be able to explain it for it to be valuable to you personally. I've learned a lot of really strange things about myself and the universe around me, Phillip, especially since I've been contemplating the powers of darkness and all. In fact, I've even traveled back to visit my prenatal existence with the benefit of this really interesting book." Rodney showed me L. Ron Hubbard's *Dianetics: The Modern Science of Mental Health*. "Because of this interesting journey into my past, I've learned that the very first face I ever knew was Satan's face. I saw him while I was growing in Ethel's womb. I know it's kind of a disgusting thought just thinking Ethel *has* a womb and all, but there you have it. Satan singled me out even when I was just a batch of simmering molecules. I guess that's why I've always been a rather unpleasant sort of person. It's not like I ever *wanted* to be such a pain in the ass—I just couldn't help myself. It was sort of like my destiny, in a way. What's really great about this scientology stuff I've been reading—this idea that we have all these infinite previous existences and all—is that it doesn't *matter*. I mean, it doesn't really matter at all who I am *now*, that I may be a devil worshiper or even worse. Because I might have been a bunch of really *nice* people in my previous incarnations. Priests and ministers, even. Kings and queens, paupers and dogs. I might have been Sir Francis Drake in a former life. I might even have been Willie Mays."

"Willie Mays isn't dead," I said.

"Yet," Rodney reminded me.

As I tapped chemicals into beakers, flasks and test tubes, as I scraped pungent growths from the surfaces of petri dishes and damp Wonderbread, Rodney often chattered animatedly like this, filling his own room with strange new notions and imaginings. There was something fecund about Rodney now, vigorous and irrepressible. Rodney lived and thought and ate and dreamed. He was a different Rodney from the one I had known before, and I must admit that now I liked him a lot better. He seemed more involved in life. He didn't just wait for things to happen to him.

One afternoon when I arrived at his room Rodney had erected the collapsible card table and covered it with a somber damask tablecloth. At the center of the tablecloth a white candle burned auspiciously on a white plate. A few of Rodney's books—*Demonology and the Occult* and *Making the Spiritual World Work for YOU*—lay haphazardly about on the floor. Three chairs were situated around the table, and Beatrice sat in one of them.

"Hello, Phillip."

I turned to Rodney. "What's she doing here?" I asked.

"The spirit world is a very feminine place," Rodney said, motioning me to sit down. "It's filled with all sorts of feminine forces or something. We need her to help us reach into the feminine half of the void. Now, does anybody want a Coke before I turn off the lights?"

Beatrice did. Then the room went dark and we all joined hands around the table.

"All right," Rodney said. "I guess we can get started." He shifted in his seat a few times. A large poster of Aerosmith gazed down meaningfully from Rodney's wall. The luminous dial of Rodney's private phone seemed to hum faintly, and Rodney cleared his throat. "So I guess, you know, we're all gathered here to talk to the spirit world. So I guess we should sit real quiet for a moment, and just listen."

The luminous dial hummed, the large poster gazed. I could still smell the odor of marijuana and cigarettes which Beatrice and Rodney must have shared before my arrival.

"So," Rodney said, clearing his throat and fidgeting impatiently in his rattling folding chair, "I guess this is sort of the part where we have to get in touch with the cosmic vibrations and all. We're very spiritual people here, waiting to meet some interesting people out there in the spirit world. I mean, if anybody out there's listening, we're looking for a sort of guide, some sort of friendly spirit who's still tied to the material world in some way, but basically who's dead already."

I heard a throat lozenge clicking about in Beatrice's mouth. Her hand gave mine a slight, conciliatory squeeze, but she wasn't looking at me.

"So we'll just wait here," Rodney said after a few moments. "We won't say a word or disturb any of you. We just want to sort of know what you're all thinking, and if any of our former loved ones are out there, and if there's anything we can do for you down here on earth. That sort of thing. You know, like maybe we'll scratch your back, you'll scratch ours. Then if later like maybe *we* need anything from *you*, you know. Well, I think you get the picture. Now, I don't want to be hassling you and everything, so I'm going to shut up for a while, and just get in tune with your vibrations. OK? So none of us are going to say anything for a while. I really mean it this time."

I don't know how long we sat there, but it seemed like hours. Eventually the noisy lozenge dissolved in Beatrice's mouth—like one of her erotic promises, I thought—and when I peeked covertly out of the corner of my eye I saw she had fallen asleep. She sat slumped forward slightly in her chair, her tiny pink tongue extruding from her too-thin lips, like the tongue of some sleeping terrier. I wondered if Beatrice dreamed of her long abandonment of me, and if she dreamed it without remorse. Rodney, on the other hand, sat unreflectively alert in his chair, his wrists braced against the flimsy table. His face didn't flinch, nor his expression waver. I had never seen Rodney so firmly involved in anything before. As my half-open eyes peered into the dark corner of Rodney's room, I thought I saw something beginning to cohere. It was red, and hot, and tiny, like a tiny glowing red eye, a canny wolfish red eye. There was a thin ribbon of steam rising from it. Then I detected the odor of sandalwood, and recognized the cone of burning incense in a delicate brass tray on Rodney's bureau. I wondered then if it mattered, whether a vision had to make itself real in order to achieve spiritual validity, or whether the world's mundane objects could be significant too, like St. Augustine's rotting fruit.

To prevent my mind from drifting, I tried to concentrate on the spirit world and the slow, bitter ghosts of the dead and unborn I expected to find there. My active mind, however, kept returning to Beatrice's damp, muggy palm I held so carefully in my left hand. This warmth had once comforted and consoled me, like heat or light. There was a special spiritual electricity here which I had considered then private and inexpressible, but which I now considered diffuse, conglomerate and altogether human, like Southern California Edison or the Department of Water and Power. This warmth could sleep and live without me, without even thinking of me. It could be my warmth, but then someday it could be somebody else's warmth too. Somebody else's hand might hold it, somebody else might kiss it with their lips. It could go away from me and live in its own house. Warmth was a sort of spiritual force too, I realized then. Ghosts often exist even in the bodies of people we love.

Rodney began to hum. A low, Gregorian, extensive sort of hum which expanded in the room like warm air or rumors. Until now I had been squinting, but as my eyes began to relax I closed them. The darkness under my eyelids was slightly phosphorescent. With my eyes closed and my mind alert, the world made a lot more sense. Warmth, light, motion, mass, gravity, weight, space and sound. These were all around me, but sometimes I could not see or sense them because the world got in my way, sometimes even the thickness and delirium of my own body. Rodney hummed, and Beatrice's soft hand nestled in mine like some submarine creature, convoluted and brainless, a mass of uncomprehending nerve and muscle. I could travel away through worlds of weight and sound, but only with this sleeping hand to guide me back again. The world of sensation was very dim, and Rodney's humming voice trembled everywhere like loose wallpaper. Dark shapes turned around me, and I descended through notched cavernous chambers of impacted weight and mass. Sound resided everywhere, but mass resided only in strategic places, places where it waited to influence human events. Light always resided somewhere other than where it was. I traveled without body or form. I was just an envelope of heat and sensation, diffused by the radiant warmth of other hands and bodies. You couldn't make out faces or landscapes down here. You could only detect the irreducible heart of things, things like light, and motion, and weight, and mass, and sound.

"Hey!"

Someone shook my shoulder. I opened my eyes.

"Hey, did you feel anything? I think I started to feel something."

All the lights were on. Rodney stood over me, a freshly lit cigarette smoldering in his hand. He had taken off his T-shirt to reveal the green tattoo of a dragon uncurling around his pale navel. Rodney's chest was smooth, muscled and hairless.

"Didn't you feel something there at the end? Not a voice, exactly. It was like we were slipping, like we were getting through somewhere."

Beatrice sat on the edge of Rodney's bed, her legs primly crossed, turning the pages of *Biker's World Magazine* and snapping her gum.

"I don't know," I said. Rodney's room seemed very smoky and unfamiliar. I reached for one of Rodney's cigarettes from the card table. "I can't really say yet."

"You'll see," Rodney said. "We'll do it tomorrow. We'll keep doing it until we get it right." Rodney went and sat on the bed next to Beatrice. He put one arm around her. "How you doing, baby?"

188

I placed my hands in the pockets of my Levi's jacket. Concealed there were the chemicals I had wrapped in baggy sandwich bags. I couldn't even remember what they were anymore. Just chemicals, I thought.

I asked Beatrice if she wanted me to walk her home.

"That's OK, Phillip." She didn't take her eyes off the turning pages in her magazine. "I think I'll stay here with Rodney a little longer."

Rodney didn't look at me either. His arm remained awkwardly draped around Beatrice's neck. Rodney and Beatrice posed there like lovers grown bored of both themselves and friendly cameras. They had moved out of the world of motion where I once adored to watch them kiss and pet. They had entered the realm of family photographs now, but it was not my family. It wasn't even really my camera.

I let myself out the front door without encountering Ethel. The light was still on in her room, however, yellow and very bright, and I could see it from across the street, where I stood for almost an hour, wondering how cold she was in there and whether she was listening, as I always suspected, for every movement Rodney made, for every sound and every breath her cautious silence might elicit. The light in Ethel's room was hard, like space, not airy at all. I stood on the street corner near a flickering, buzzing lamppost. The light was already out in Rodney's room, and I figured they must know I was out here. It seemed to me implicit in our relationship, that I would be standing on a street corner watching the dark window of Rodney's room while Ethel's light burned cold and useless alongside it.

189

After a while, Ethel's light went out too, and, abandoned even by the abstract movement of contrasts, I returned again to my real home.

Dad was lying on the living room sofa with the television on. Johnny Carson was saying, "I don't know about you, Doc, but I'll be damned if I'll spend the night with Ed in a urinal, mirrored ceilings or not!" Dad chuckled feebly. One hand was propped behind his head, the other pressed against his stomach. An open blue glass bottle of Maalox stood beside Dad on the mahogany coffee table. As I stepped slowly into the room, my reflection crossed the Panasonic's dusty screen.

Dad cocked his head to see me. Then he looked back at the television. "Hi, son. I didn't hear you come in." Dad took a swig from his Maalox and replaced it on the coffee table, which showed a few faint white rings where Dad had placed the bottle before. "I think I'll sleep in here again tonight," Dad said. "I've been restless lately. I don't want to wake your mom or the baby."

"Do you want a blanket?"

"Sure, maybe before you go to bed. How were your friends?"

"OK, I guess."

"You should bring them by sometime. I'd like to meet them."

I was staring at the television. "I'm sure they'd like to meet you too," I said emptily, without any enthusiasm. Johnny was talking about Ed's drinking. Johnny didn't want

190

to say Ed drank a lot but. Johnny'd been out with Ed a few times, and wow-a-wow-a-wowa.

"How's your stomach?" I asked Dad.

"I don't know. I think it may be getting a little better."

I fixed Dad strong black coffee laced with tannic acid, brought him a blanket and a pillow from our well-stocked linen closet (stocked, incidentally, by an invisible maid who always arrived and departed before I was out of bed in the morning). Then I went down the hall into Mom's room, where Mom was watching Johnny Carson too.

Mom was radiant, propped up by a brown corduroy backrest. On a handsome tray beside her bed was an ice chest, bottles of Perrier, orange juice and a small untouched bell glass of rosy port.

"Hi, Mom," I said.

I could see the flickering television reflected in Mom's eyes. Mom's protuberant stomach underneath the taut, tucked blanket was filled with movement, just like Dad's. Mom's movement, though, was the movement of life. Molecules assembled, deploying minerals, proteins and enzymes. Routine circulatory, digestive and pulmonary processes began to beat inside the still gelid mass of it. You could already begin to see the tiny eyes. You could already begin to hear the tiny mind learning to click, click.

"Dad's sleeping on the couch tonight," I said, crawling onto the bed beside her, touching her blanketed stomach. "So he won't wake you or the baby."

"That's nice, dear."

"I want you to know I'm taking full responsibility for

191

everything that happens from now, on." I had taken Mom's fair freckled hand between my hands. The skin was soft and faintly translucent, knitted with fine blue veins. "I'm not going to go drifting off again when things get too confusing or complicated. I know I went pretty far into myself for a while there, but I've come out the other side, now. I think I've grown up, Mom. That means I can be a lot more help to you from now on."

Mom transferred the glass of icy Perrier to her right hand. Her left hand touched my wrist. For the first time since Dad's arrival, Mom was wearing her expensive wedding ring and engagement band which used to lie neglected in the bottom of her big black purse among crumpled gum wrappers and Kleenex. "I'm sure you will be, baby. I'm sure you'll be a lot of help to me and your father. You always meant well. I never thought for one minute you didn't mean well."

I fell asleep that night in Mom's bed, with the warmth of her hand on my wrist. This was the vital current. Beatrice's hand, mine, and now Mom's. It was like genealogy, race, intertribal culture, migrating birds, evolution. The warmth traveled from one person to another; it changed people and people changed it. It grew warmer, it grew colder. Sometimes it grew warmer again. Soon Dad would be a part of it too, I thought, pulling the smoky dreams into my body and face. Soon Dad would be the warmth I shared with other people and there was nothing, really, nothing he could do about it.

22

"The occult is that relative half-world into which we jour-ney to make our own world more real," Beatrice said, while Rodney and I squirted Hansen's airplane glue into our Kleenex. I kept my eye on Rodney. "I don't think the occult dimension is necessarily more invalid than our own, or more valid either. It's just that big black gap of things we don't know. I think we should learn to accept the things we don't know on their own terms, without wondering things like, you know. Whether we're *really* talking to our departed great-grandmother or not. Or whether it's the devil out there, or demons or goblins. We have to accept life's gaps and lapses as well as its hard promises. I think that's your problem, Phillip. I think you need answers to everything. I think something's not real to you unless you can use it exactly the way you want."

Rodney pinched shut his left nostril and applied the wad-ded Kleenex to his right. Then he inhaled a long whistling rough breath and the Kleenex popped. I followed his example, though perhaps with less ardor.

"I mean, if you guys could just see yourselves," Beatrice said, overturning her book on her knee. The book was Nietzsche's *The Genealogy of Morals.* "Shoving crap up your nose so you can feel less real. Trying to move into the half-lit world of the doped, the dreaming and insane. Now, if you guys were just trying to experience it, I might have a bit more sympathy for you. If you were just charting maps, trying to move to the edges of the experiential precipice, so to speak, then I'd just say you were kids stretching your wings. It would just be part of growing up—and a *good* part, as far as I'm concerned. But you guys don't think that way. You want to move into the unreal so you can turn it into property. You want to build houses there, motels and swimming pools, convenience stores and parking lots. You want to find escape, pleasure, pain, spirits, things. Things you can use, just like you use those chemicals, Phillip. You don't care about chemistry. You don't care about abstract knowledge—no matter what you say. You're just using those chemicals to kill your dad. And that's what this whole occult thing is all about, too. So you and Rodney can take control of everything, because that's how you see life. Use or be used. That's so goddamn male, Phillip. That's so goddamn hopelessly . . . oh, I don't know. So goddamn *penile* of you. It makes me just want to throw up. You notice I'm talking to you, Phillip, and not to Rodney. That's because I always expected more from you. It's not like Rodney's ever listened

194

to a word of sense in his entire life—certainly not if it would do him one bit of good. But I always expected more from you, Phillip. I thought you and I shared a certain unexpressed sympathy about the unknown world. I thought you loved it as much as I did. But you're no better than Custer. You're no better than the goddamn Mormons. You just want to make the unknown profitable. I guess I've been really disappointed in you lately, Phillip. I guess I've really been disappointed in you ever since your dad came."

A hot burning sensation lifted high into my skull. It was not unpleasant, and quickly diminished to a soft convincing whisper. I rested the wadded Kleenex on my knee. I felt very good and relaxed. It was not as if I was under the influence of anything at all. It was a sensation just like breathing or drinking cool water. A thin fog entered Rodney's room. I heard Rodney taking another long sniff, and when I looked at him again he was withdrawing the Kleenex from his nose. Fragments of the pink tissue were attached to the rim of each nostril like confetti.

"Pretty heavy, huh?" Rodney's voice was nasal and rough. He leaned back on his bed, adjusting pillows, then crossed his arms and looked at me, grinning remotely like a forgotten relative at some tedious family get-together. "Do you dig this stuff, dude, or what?"

Beatrice was wrong. I did not feel more distant from the world so much as more firmly rooted in it. I was caught among its tangled whispering earth and dense folds, like a sparrow in a blanket. Small animals hypnotize themselves when trapped by some urgent predator. It's not so painful then, life's last moment.

"Just keep pouring more fuel on the fire," Beatrice said, as if she were reading my mind, or I were reading hers. "Keep acquiring more property. Keep buying more nice pretty things and stuffing your bodies with more burgers and cheesecakes and candy bars and bonbons. And don't stop just because you don't want any more, or even because you don't need any more. Because you always want more. You always need more."

I sniffed. Rodney sat across the table from me. His expression was remote and diffuse. He sniffed.

"She never shuts up, does she?" he said. "All day and all night. Yap yap yap. Even in bed when you're doing it to her, she talks right through it. She doesn't shut up for one minute." Rodney sniffed, and ran the back of his hand across his nose. His eyes were red and bleary. "It can get pretty frustrating sometimes."

We all joined hands around the table again, Beatrice with her armed and petulant silence, more ominous even than her most dire predictions. Rodney and I shared our gazes with the flickering candle flame.

"Now, I've been reading up a little more on this stuff," Rodney said.

I sniffed (or was it Rodney who sniffed?).

"Now what I should've done first is gone off and purified myself, by drinking pure water and meditating," Rodney said. "But since we didn't do it last time, I figured we might as well skip it this time, too."

"You'll both end up in reform school," Beatrice said. "You think you're a couple of real rebels or something. But when I look at you all I see are a couple of little kids."

Rodney sniffed noisily. After a moment, I sniffed too, but more succinctly, as if to dissociate myself from Rodney.

"Now," Rodney said, switching off the light and taking our hands, "let's get down to business."

Over succeeding days and weeks the three of us settled into our new routine with a sort of grateful complacency. It was like the days of our burglaries, only more patient and informal. I would spend the day at home watching TV with Mom on the big bed, while Dad was either at work or sleeping fitfully on the living room sofa. If Dad was home, I would hear him go into the bathroom every twenty minutes or so, then the flush of the toilet. Mom and I watched the game shows and soap operas together, and sometimes I tried to explain the rationale of these programs to Mom's baby.

"The woman's voice you're hearing now," I said, "is the voice of Victoria Morgan, the youngest and very spoiled daughter of Joshua Saner Morgan, the richest man in Creek Valley. Victoria is used to getting her own way, and she'll get it no matter who tries to oppose her. She's not really the mother figure in all this. She's sort of the evil-twin type, though she doesn't look at all like her sister, Felicia Morgan, who's actually a very nice person. Felicia's daughter, Jeremy, is the illegitimate daughter of the police inspector, David Rampart. I think of her as the sort of heroine of the

197

show, because she's really pretty—almost as pretty as Mom— and she's always trying to help her friends get out of trouble, like when they tried to frame Tad Stevens for murder that time. When you get out, we'll probably watch "Heartbeat County General" every afternoon, so eventually you'll catch on to all the names and faces, though you don't really need to know all the characters and plots that well to enjoy the show. Also, I better warn you. People are always extra se- rious about how they feel on this show. They're either *really* happy, or *really* sad, or *really* having a good time. It can really get on your nerves after a while. It's something you just have to get used to, I guess."

Late in the afternoons I would tuck Mom into bed after she had fallen into one of her dozes, then slip down the back stairs to avoid Dad on the living room sofa, and arrive at Rodney's around four o'clock. Ethel would present me a small snack. We would exchange some light banter about weather and current events. If I waited long enough, she would inevitably try to engage me in conversation about Rodney.

"Is there anything bothering Rodney that you know about?" she might ask, while I munched my carrot sticks, or contemplated my hot black coffee. "Is there anything Rodney's not telling me?"

I tried to remain as noncommittal as possible, but con- fronted by Ethel's strained and withering sadness I could not help but offer her tiny gifts.

"I think it's just school," I might say. "He's having trouble adjusting." Or: "I think it's just his age, Ethel. You know he's practically a teenager and all. Everybody starts acting a little weird when they go through puberty, or so I've heard."

"I don't know where he goes at nights," Ethel said. "People call on the phone for him. Sometimes they're grown men. I don't trust their voices. They have very dark, rough voices. They want Roddy to meet them places. Some nights, Roddy doesn't come back at all from these meetings. Sometimes he's gone for days at a time, and won't tell me a word about where he's been. He won't even call to tell me when he'll be back."

"He's just practicing a little independence, Ethel, that's all. It's perfectly normal for kids his age. Especially for young men."

"When he's in his room alone at night I hear him talking to himself out loud, saying the strangest things."

"He's just exercising his imagination. Rodney," I assured Ethel, as if it were some sort of cherished compensation, "has a very active imagination, as you well know."

"Sometimes I get very worried, Phillip. I can't sleep at night. I start thinking, well . . . I start thinking terrible things about Roddy. I'm embarrassed to admit it. But I start thinking maybe he's not turning into a very nice person. I can't help myself. I know it's a horrible thing to say . . ." Ethel turned and looked away. Tears formed in her eyes. Her voice grew rough and swollen. "I just don't know sometimes." One tear ran down her cheek.

"It's OK, Ethel. I understand." I reached across the table and took her hand. "I really do understand, Ethel. Don't cry. Sometimes we hate the people we love. Freud said that. It's perfectly normal. Whatever you think, for God's sake don't think there's anything wrong with you, because there isn't anything wrong. There really isn't."

"I don't mean hate." Ethel took her hand away. She had abruptly stopped crying. Her voice had returned to normal. She gazed off at chrome fixtures on the stove, her hands nestled together in her lap like lovers. "I mean I start having these terrible thoughts. Roddy's a *good* boy, and I know that. I really do. But sometimes I think, well, he may have gotten in with the wrong crowd. I don't mean you, Phillip, or Beatrice. I mean the sort of crowd he hangs around with when you're not around. Like the boys at the bowling alley, or the boys who hang around at Shakey's Pizza. Then there's those strange boys I always see outside on the street at night. There's usually just one of them. I can see him from my bedroom window. Sometimes he's staring up at me, like he knows I'm watching. They frighten me, Phillip. I'm so worried about Roddy I can't sleep or go to the bathroom I'm so worried. I think it's happened. You know, Phillip, I think Roddy may be doing it. He may be experimenting, you know. With drugs, marijuana. Or maybe even worse."

"Ethel, look at me." My voice was very firm and direct. I leaned earnestly across the table, pushing aside a depleted wooden bowl of corn chip fragments. "Ethel, look me in the eye." Her hands began to fidget as her eyes rested on mine. The hearts of her eyes were fractured and simian. Her hands began pinching at one another, like quarrelsome crabs. "Rodney has a cigarette every once in a while. Maybe a beer or a drink. But he's not crazy, Ethel. He's got his feet planted firmly on the ground, and you *know* that. I'm Rodney's best friend, and *I* know that. OK, so he likes to wear some offbeat clothes—but that's Rodney. He's a trend-setter, he's his own man. I'm telling you—all the kids at school look up to Rodney. They're always imitating him. If Rodney's going to keep one step ahead of the pack, he's got to go for the more outrageous sorts of styles, you know? But, Ethel, now listen to me carefully now. Don't ever, *ever* think *that* about him again. It's just plain wrong, that's what it is. It doesn't do you any good. It doesn't do Rodney any good. Just trust him, and trust me. Rodney's a good kid, Ethel, and you know that. He's good inside—he's good in-side *here*"—I thumped my chest affirmatively with my fist— "in here where it matters. He's going to turn into the sort of man you're going to be very proud of one day, Ethel. I promise you."

I got up from the table, took the bowl to the sink and rinsed it in cold water.

"You're right, Phillip. You're right. I shouldn't doubt him. You're right, Phillip. Oh, you're so, so right." Her eyes were filling with tears and admiration for me, even longing. Our conversations always ended this way, with her palpable de-sire for me filling her eyes with tears. She wanted to keep me here, I could tell. She wanted to keep me here forever, her adoring and adored second child. She wanted to lift me up and hold me in her arms as tightly as any lover. I couldn't stand it—I finally understood how Rodney felt. I had to get away from here. I had to get far, far away. I had to get as far away as I possibly could from Ethel's oppressive arms.

LIFE

23

After our little meals Ethel would grant me a few of her Ziploc sandwich bags and I would go upstairs where Rodney was listening to Judas Priest or vintage Alice Cooper on his ghettoblaster. Removing his chemistry set from the closet I would check my pocket notebook and mix new, untried combinations which I then wrapped in the plastic bags and hid in my inside jacket pocket. Then Rodney and I would smoke a little marijuana or a fragment of hash, just to put ourselves in the mood. If we were feeling particularly uneasy or discomposed, Rodney would get the airplane glue out from under his bed. By this time Beatrice had arrived, presenting us with baleful forecasts of our adult years when we would certainly turn out to be just like our fathers, oppressing women with our corporations and pocket calculators, adding to the world's heartless mountains of wealth, credit-wealth, consumer goods and other mere things and numbers. "You're all the same," Beatrice complained, reluctantly taking our hands at the dilapidated card table, which

was chipped, water-stained, and tracked in places by globu-
lar wax. "You just want to be the center of attention all the
time. Me me me. That's all you care about." I was already
feeling tremulous and thin from the grass and the glue. My
eyes felt slightly sore and fuzzy. But I felt lucid as well, lucid
inside my own mind, where hard crystal shapes emerged,
and spirits gathered firmness, gravity and substance. These
were real things in here, not just ideas or shadows. When
I closed my eyes everything suddenly made more sense.
Then Rodney would begin to hum.

On a Wednesday, separately and silently in Rodney's dark
room, we all experienced our first real encounter with the
purer world of spirit.

I was feeling particularly heady and diffuse with marijuana and glue, sitting on my seat like a swami floating on
a carpet. Lights, candle, smoky incense, cabala, grimoires,
totem and taboo. "I think we've been patient long enough,"
Rodney said. His entire tone and demeanor had changed.
He was wearing a Nazi insignia framed by a pentagram on
a leather thong around his neck. The knuckles of his right
hand displayed his new tattoo, a bright caduceus with rip-
pling scaly skin. His hair was shaved in a mohawk and dyed
an almost fluorescent orange. "We're not looking for hand-
outs, you know. It's not like we're asking for favors or anything. We just thought, like, we're young, and we're going
to be around on this earth for quite a while, and we're
willing to do literally *anything* you want, and all you've got
to do is just *speak* to us for about five seconds and let us
know you're interested. But do you guys have the time? I

mean, do you guys even make the measliest little effort? I don't think you seem to appreciate all the time and energy my friends and I have been wasting here."

Beatrice took her hand from mine momentarily and yawned, holding her tiny hand to her smudged moist mouth, her eyes closed while she stretched. She looked like a small white kitten when she yawned. Whenever Beatrice yawned it reminded me of how much I once loved her.

"So anyway," Rodney concluded, "let's get this relationship going, and stop beating around the bush. We're here, you're there and it's about time somebody made the first move. You guys are more experienced in all this than we are—we're looking for a little mature guidance in the matter. So let's *go*. Give us a *sign*, for chrissakes. We're starting to look pretty stupid, if you want to know the truth. Holding hands and waiting for you to take your sweet time and all. Look, you want sacrifices, blood rituals? You want our *souls*, for chrissakes? I'm not doing nothing with mine—it's *yours*. You hear me? Come and take *all* our souls—all except Beatrice's, of course, because she's such a perfect angel, as everybody knows in the entire universe by now since she's probably told them herself personally. Phillip and I, on the other hand, don't give a fuck. We're yours. But if there's something you want us to do, then you've got to tell us. OK? Are we getting through to you guys? Now, we're going to be quiet again for a while, but that doesn't mean we're like suckers or something. That doesn't mean we're going to sit here forever. I'm sorry to be sounding so impatient and all, but I'm starting to feel a little used, if you want to know the goddamn truth. So look, whenever you're ready, you just let us know, OK? We'll sit here nice and quiet, and you take your time and think about it. Then, if you want, you contact *us*, OK?''

So we sat in the darkness, and Rodney hummed, and Beatrice fell asleep, snoring slightly. It seemed a night like all the others, chemicals in my pocket, this strange house of Rodney's around me, Ethel downstairs with her doubts and unsteady aluminum cane, while back at home Dad was being steadily dissolved by the universe of rushing darkness and Mom watched color TV. Again I descended through the earth's dark layers into a subterranean world where strange prehistoric skeletons etched the dense basalt walls; broken human bones and teeth lay strewn about like discarded toys in a cannibals' kindergarten. I was expecting to find the dead down here, spirits with scores to settle, or even vast shapeless things without thoughts, things that just shifted and turned. Perhaps it wasn't the afterlife at all. Perhaps it was the pre-life. Or perhaps it was just nothing and nowhere, where beings just lay around waiting for things that never happened. Non-life, anti-life. Proto-death, death in life. I was moving through a convoluted passage which seemed only dimly familiar. Death. I'd never encountered it before, not even in my imagination. Death was in these passages I had until now blithely elided in both my texts and my dreams. Death was something permanent. Death never moved. I began to feel an ominous presence in the darkness around me. Dead. Voiceless. Pitiless. Lucid. Hard. Death was matter, death was pure mass. Death might even be better, I was beginning to suspect. Death was real, while life by contrast seemed little more than a presumption, something broken which rattled and would not last. I was moving into the lightless heart of something I had never seen. It was filled with shapes and presences, but you could not see or touch them. Instead they seemed to elicit a sort of buried radar from my skull and sinuses and teeth. The world of death was very simple. There was no more thinking or being thought down here. There was no more fear or suffering or hate. Ever since I could remember I had been trying to discover my own Way in life, that journey I would make in the world alone. Perhaps I had been looking in the wrong place all along. Perhaps this was my true path down here. Something cold passed by me. Everything was growing misty and damp as I waded into the mud which grew deeper and marshier. It grasped my ankles, calves, thighs. It made sucking sounds against my skin. I realized I wasn't wearing any clothes. I felt very cold all of a sudden, as if all the cold shapes passing me in the cavern were now gathering around, pressing closer against me, untextured and weightless and dull.

"It's all the same slow dream," Mom's voice said out loud, loud and real in this underworld like the voice of the Mom who, I understood only now, was dead forever. "You, me, Dad, our home back in Bel Air. It's a beautiful big house that's waiting for us, baby. It has a pool and a big yard."

"This is the history of motion," I said. "You and me, Mom. The history of motion."

"There's nothing for you down here, Phillip."

"I want to stay."

"You wouldn't like it. You'd catch cold. I wouldn't always be around, and you'd be frightened. You'd wonder where I was. After a while, you'd start to resent me."

"I resent you now."

"You've spent too much of your life alone, baby. That's my fault. I never helped you become properly acclimated to the world. There's a real world in which we all have to live together. That means we have to make concessions for the benefit of other people. That means we simply can't have everything exactly the way we want it all the time. This life you're living inside yourself is just a dream. A dream of you, me, and your father which doesn't work. Or maybe it works too well."

"I've decided I'm going to do it." I could not disguise the lift of triumph in my voice. "Rodney's going to help. Rodney and I are going to do it together."

"Then do it, baby. At least you'll be functioning. At least you'll be making some sort of impression out there, instead of just down here in your own mind. Live in the world, baby. That's all I ever meant for you to learn, and you never did. It's my fault. I can't blame you. It's my fault entirely."

"Is Pedro down here?" I asked. I had felt another cold shape approach me. I was filled with either fear or hatred. Something burned in me, some impalpable fuel. "Is that Pedro with you?"

"Of course not, baby. Pedro's upstairs with you. Pedro's out there in the real world with you."

210

"Pedro?" Tears were forming in my eyes, cold, freezing. "Pedro? Is that you? Pedro? Are you out there? This is me. This is Phillip."

"If you're going to do it," Mom's voice said, "then you better do it now. Stop beating around the bush, baby. If you're going to kill your father, then kill him tonight. Kill him tonight and get on with your life."

It was Pedro out there, but Mom was hiding him from me. It was Pedro. His multitudinous arms came up around me, icy and damp and formless and thick. It was Pedro. Pedro was dead and waiting for me. For me. Pedro was waiting for me.

"Why don't you go kill everybody while you're at it? But start with the men. Kill all the men, you guys. Then get back to me in a few thousand years or so. We might have something going then. We might be on our way towards some practical solution to things."

"If you're not going to help, Beatrice, just put a lid on it, OK?" Rodney was selecting a cord of rope from a cabinet drawer. We were standing in Rodney's basement in the cold fluid light of a naked overhead bulb. I was on my knees going through my steel toolbox. The lid of the case was open, displaying cold gleaming tools in red steel compartments.

"I'll tell you what. *I'll* even help." Beatrice was sitting atop Ethel's Maytag dryer. She had lit a cigarette, and was gesturing it with dramatic disregard just like Greta Garbo. "We'll buy us some machine guns, some Uzis, and some

211

army issue bazookas. Then we'll go down to the shopping mall and start blowing everybody's goddamn head off. OK? That'll teach them. That'll teach them all a good lesson or two, won't it? I can hardly wait. I really can't wait another minute."

"Do we need these?"

Rodney showed me a rusty pair of gardening clippers.

I thought for a moment. "Sure," I said. "Why not."

"Look, we'll get us some hand grenades. We'll start lobbing these big hand grenades into B. Dalton's and the May Co. We'll blow the fuckers to kingdom come—that's what we'll do. We'll be just like Dirty Harry. We'll be just like John Wayne. Sure, some innocent lives may be lost, but there's nothing we can do about *that*. If you're going to fight evil, ma'am, then sometimes you just gotta be a little evil yourself. We'll detonate the goddamn mall, that's what we'll do. Save ourselves the trouble of going in there. Then we can move on to the Ford dealership. Cost Plus. The Warehouse. City Hall. We'll teach all those liberal phonies what real suffering's all about, won't we, guys? Sure, they can all *talk* about peace and love and brotherhood, but when it comes right down to getting things done, well, that's where *we'll* move in. Whistling our national anthem and spraying bloody death wherever we go, because we're *realists*. We want peace too, but we don't have any liberal bullshit illusions about how it's gotta be achieved. War's hell, but sometimes it's just goddamn necessary if peace is to be preserved. I'll meet you guys in the car. I'm going to go take a long piss in the alley." Beatrice flicked her cigarette and it arced across the dim garage, crashing against Ethel's Toyota Corolla in a shower of sparks. "Bring beer. Afterwards we'll go down to the whorehouse and get ourselves properly laid." Beatrice stared at my toolbox with a sort of weird inanition.

Rodney stacked a long extension cord beside the toolbox, in case the rope was not enough. We stood there for a moment thinking, looking at the gray steel box.

"Next stop, the Middle East," Beatrice said. Her voice was clipped and mechanical. "Peace through strength. Wealth through poverty. Love through death. Once we've taught those Palestinians a lesson they'll never forget, we'll build this humongous K Mart. Then we'll move on to take care of those fucking Chinese. We won't even try to set up any sort of provisional government there or anything. We'll just kill all the fucking Chinese. God, how I hate them. God how I hate those goddamn Chinese."

Rodney's hands rested on his hips. His expression seemed momentarily to approve of our preparations. Then, as if approval in any form was for Rodney a sort of lapse, he scowled bitterly. He reached to his shirt pocket for his cigarettes, offered me one, and shrugged in Beatrice's general direction.

"She thinks this is all some sort of game," he said. "She thinks this is all just some big har-de-har laugh or something."

"I know," I said. It was time for us to leave. "But in her own way, I think she's trying to understand, Rodney. We've got to give her credit for that much." Then I knelt and cranked shut the toolbox's strong steel clasps.

24

"We've unleashed strange forces in the world tonight," Rodney said. "That's what confuses her. In fact, that's what I think women in general don't ever seem to understand. Not that these forces exist. But that we can use them. They aren't just ideas. They accomplish things. They go places."

We were laboring down the ill-lit rubbishy streets off Van Nuys Boulevard, Rodney carrying the toolbox and I the cords of rope, electrical extension cords and a few random saws and hammers. The night air was thick with smog, palpable and rough. Like the smog itself, the darkness did not radiate so much as settle over everything.

"That's the illusion women prefer. That everything can be reduced to talk. I'm warning you, Phillip, I got this one figured. Women are really great and everything. I'm not saying otherwise. But they've got their own sort of truth and it has a way of confusing things sometimes. Men do things. They get things done. That's what men do. Women, on the other hand, talk about things. Why they weren't done quite right. How you might want to go about doing it better *next* time. Which things to do first, and which things last, and which things after that. Talk talk talk. I mean, like Beatrice and all her Communist bullshit. She wants to feed the world, right? But I don't see her feeding anybody. I mean, when's the last time you ever saw Beatrice feeding anybody? I'm talking even a sandwich or something. Never, that's the last time. But when's the last time you heard her *talking* about feeding everybody? She's got a million ideas as far as talk's concerned. If talk was wheat, Beatrice and her Communist sympathizers could feed the whole world. But talk ain't wheat. It's nothing like it." Rodney's pace and expression were gripped by sudden purpose, as if he and I were hurrying to impart some crucial theorem to Rodney's bemused colleagues back at the lab.

"These forces we've unleashed tonight aren't new things, Phillip," Rodney explained earnestly, shifting the toolbox into his left hand, swinging the entire weight and balance of his body along with it. "These forces have been around forever and ever, since the beginning of time, in fact. They've been around since before mankind, even since before the dinosaurs. They were roaming around the universe before the Earth was anything more than a bunch of cosmic debris. They knew what they wanted the world to be, and so they made it that way. They didn't talk about it; they *did* something. They got things done."

Needless to say I was elated, higher than a kite, breezing through the muggy breezeless night. In the wide sun-bleached and pitted streets we walked past dilapidated au-

tomobiles, fading lawns and houses, thinning and recalcitrant trees and foliage. Mimosa, jacaranda, fig, palm,
eucalyptus, dry and spotty bamboo. The air was pungent
with gasoline, smog, and the fishy smells of cooking which,
along with brief bursts of salsa music and Julio Iglesias, filtered from the expressionless facades of Latin households.
Inside those houses people glanced out fitfully from behind
cracked venetian blinds. Timid small children with big eyes
hid behind their parents' legs, waiting for their mothers to
drive them to the laundromat, supermarket and home again
in broken automobiles. At supper they ate with vaguely
surreptitious expressions, their ears alert for any sound in
the street, awaiting that penultimate knock on their door.
These were families who were always waiting to be sent
away, and as a result you never really saw them. These were
the citizens of my secret community I most cherished and
admired. They, like me, lived their secret lives in public
places.

I was going home and taking Rodney with me this time,
and that made a difference. My dreams weren't secrets anymore, but rather part of a common purpose, a scheme of
shared knowledge. Rodney and I were going home together
to see my dad.

Rodney was right—we had unleashed strange forces tonight. Severe black things that had moved up from the
caverns of the dead. Every once in a while I felt them bus-
tling invisibly past me in the street, obloid and featureless,
like faintly disembodied laundry hampers. You could hear
tires careening on Sepulveda Boulevard almost a mile away.

Everything in the world seemed to be aligning itself with these invisible forces, assembling like military units or pieces in an intricate, vast puzzle.

"I don't think evil's such a bad thing, really," Rodney said. "I think it's just something we've got to get used to. It's certainly been around a lot longer than we have. It'll certainly be here long after we're gone. During our séance tonight I heard it speaking to me, Phillip. It said, Get on with it. Live your life. Make things happen. If you listen to Beatrice, you'll never go anywhere, you'll never do anything. For some reason, when I heard that voice, I wasn't really excited or anything. I felt sort of bored, really, like everything had already been figured out. It wasn't something I really enjoyed, just something I simply had to get over with. Frankly, that's what it was like the first time I went to bed with Beatrice. It was like I had to get this over with. I really couldn't get into it that much, once we'd started and I was getting the hang of it. It was like mowing the lawn, or putting away groceries. I guess it's because I'm evil, Phillip. Ethel told me that once when she was really angry. Even a teacher once—even a teacher told me once I was evil. That I'm no good, a bad seed, a black sheep. Evil is just mechanical activity, Phillip. That's what I think, anyway. There's no thought behind it at all. It just goes on and on and on. They say the universe started from a tiny ball of matter, no bigger than this toolbox. I always thought the universe would be a lot more various than that, but now I see it's just the same stuff, stretched out all over space and eternity, filling everything."

217

We had arrived outside my house. I felt the rushing, formless shapes hurrying faster around me in every direction. There was no wind, no sound. In the living room window, the light was on. The colorless heat of the television glowed steadily.

"I guess you can probably tell I've never really been that big a fan of women's lib," Rodney said. "I say let women stay at home and talk all they want to. Men are the ones that get things done." He wasn't looking at me. He was looking at the toolbox, which he rested against his knee. The ropes and cords were wrapped around my neck like the bandoliers of some South American revolutionary. Rodney was right. I felt bored, inanimate, sleepy. Quietly I opened the garage door, and we carried all our materials up the basement staircase. Mom would be asleep by now. If we hurried, we could be finished by morning.

25

Dad hardly stirred on the sofa when Rodney and I entered the living room. I closed the back door and turned the dead bolt. Both the radio and television were on. On the radio Rosemary Clooney was singing:

So kiss me once, then kiss me twice,
Then kiss me once again,
It's been a long long time . . .

Dad lay contentedly asleep, one arm across his chest, his head tilted to one side. The bottles of Maalox stood empty on the coffee table. Dad seemed posed and forlorn, like an expired romantic youth in some pre-Raphaelite painting. On the television Tom Snyder was discussing the secret lives of "sexual deviants" with a transvestite. "Though of course if you look at it from their perspective, I guess," Tom said, "it's probably the rest of us presumably *normal* people who

219

seem like the *deviants*. I mean, we all do our own thing, right, and then when somebody doesn't like us they say that we're a deviant. But for want of a better term, and since such people are often linked in our minds with the term deviant, I guess I'll rudely refer to our next guest as just that, and hope they can understand and bear with me for just a little while . . ." Tom Snyder chain-smoked and gestured vigorously at the camera with both hands. The close-up of Tom's head framed by his easy chair made it appear as if he were in the living room with us. The transvestite's back was to the camera, and his voice was being distorted by the sound engineer. "It's like waking up every day knowing somebody will find out," the transvestite said. "Somebody very close to you. Somebody who loves you, and believes she knows you, and yet she doesn't really know you at all."

> *You'll never know how many dreams*
>   *I dreamed about you.*
> *Or just how empty they all seemed without you*
> *So kiss me once, then kiss me twice*
> *Then kiss me once again,*
> *It's been a long, long time . . .*

"Should we wake him up?" Rodney asked.

Dad's lap-top, his briefcase, his stacks of printout and papers were on the dining room table. Some coffee in a mug splotched with scummy cream. A moldering and half-eaten French-style donut on a sheet of corrugated white paper towel.

"It's up to you," Rodney said. "He's under a spell I cast back at my house. I like to call it the Spell of the Sleeping Man. When men sleep, their souls travel around the world, trying to shape themselves into other things. I've had your dad's spirit held incommunicado. He won't wake up again until I let him."

I switched off the radio. "I want to check in on my mom," I said, and led Rodney down the hall. Mom's bedroom door was open, and Mom was sitting up in bed. Her face was brilliant with bright cosmetics and white, grainy talcum. Her hair was bundled up in a black net cap. The light from the television flickered across her face, like radar scanning the moon. She looked a little like Bette Davis in *Whatever Happened to Baby Jane.*

My hand rested on the cold doorknob. I wanted to pull it shut before Rodney saw her. But Rodney was already standing there beside me. It was important to me not to appear ashamed. Now that Rodney saw her, I wanted him to be able to take his time.

"Jesus," Rodney said after a minute. I was starting to tremble, awaiting Rodney's judgment concerning the truest, most secret part of me. "For some reason I thought your mom was good-looking."

"She used to be," I said. "She used to be really good-looking. She's started letting herself go lately. I think it's because Dad came back. Or maybe because of the baby."

221

Outside Mom's bedroom window a searchlight flashed through the alley. Out of the city's general white noise emerged the hard beating sound of a helicopter in the air. The searchlight flashed again.

"Does she just watch TV all day?"

"No." I shrugged. "She talks to me sometimes. Sometimes she talks to Dad."

"What sort of programs does she watch?"

"Old movies, usually. Game shows and soaps. That sort of thing."

"It doesn't like give you the creeps to have her sitting there all the time with that look on her face? I think it would me. I think it would really give me the creeps."

"Long ago Mom and I came to a sort of understanding," I said, and realized suddenly it was not the sort of understanding one could easily explain to a third party.

On the television the transvestite was saying, "It was nice just to meet people who understood how I felt, and didn't make me feel like some sort of weirdo or something. It was nice to know I wasn't alone, and that what I was doing was perfectly normal to a lot of perfectly normal people like myself. Many of these people had prominent careers in business, advertising and even television broadcasting."

Rodney struck a match and took a long hit off a joint. He took two more quick hits, just to get the ember flaring. Then he handed it to me, speaking deep in his chest while he held his breath.

"She's knocked up, isn't she?"

222

I looked at the bright ember. A hard green seed blackened and spilled onto the rug. I stepped on it.

"I know," I said. "She's really knocked up."

Then I pulled shut Mom's bedroom door.

We bound Dad on the sofa with the clothesline and electrical extension cords and gagged him with a pair of his own white monogrammed handkerchiefs. Dad didn't move or make a sound or open his eyes. His face was flushed and pouting, like the face of a small child who has just awakened in the lap of its parent at some endless holiday party. There was something very warm and innocent about Dad now— if I still believed in innocence, that is. Occasionally, when his head lolled to one side, he might momentarily snore or kick. Dad was definitely very far away. Perhaps Rodney really had managed to arrange his spiritual kidnap. I had a pretty good idea what we had to do now.

I was high on the marijuana and my first few sips of a Budweiser I had found in the refrigerator. Rodney was drinking Jack Daniel's on ice, crushed glittering rattly ice from the freezer's automatic icemaker. "It's not what we do that matters," Rodney said, opening the toolbox on the floor. "It's our frame of mind when we do it. This isn't another person, Phillip. The person inside your dad's already dead. This is just a body filled with energy. This is a body filled with energy that we can join ourselves with and use to make ourselves stronger. We can shape, funnel and redirect it places. We can use it to our own purposes."

Rodney handed me one of the sharper tools. "All right, Phillip? Do you understand? How do you feel? You ready to go?"

The tool felt very firm, like the edge of a desk, or the fender of an automobile. The weight was reassuring, in a way. But there was something in its dull edges that disturbed me for a reason I couldn't quite articulate. "I don't know," I said. I waited for a moment. Already I felt his presence in the room, as if a large window had opened to admit a soft cold wind. He deserved to be here, I thought. It was perfectly fair. I turned and looked over Rodney's shoulder. Pedro was sitting at the dining room table beside a stack of Dad's business papers. "Do it," he said. "Do it, do it, do it."

I looked at Rodney again just as Rodney turned to look over his shoulder. After a moment, he looked at me. He didn't look like he trusted me very much.

"I feel good," I said quickly. I didn't want him to see Pedro for the same reason I didn't want him to see Mom. It seemed to me a sort of personal violation. "I think I feel all right."

I leaned over Dad with the sharp tool in my hand.

"This has nothing to do with your dad at all," Pedro said very loudly. "I don't have a single bad word to say about him. It's your mom we're thinking about now, Phillip. I think it's about time you stopped worrying so much about your own damn self and started paying a little attention to *her* feelings. You shouldn't have done what you did to me, Phillip. I was good for your mom and you knew it. You only loved your dad because you knew he wasn't any good for her at all. Your dad didn't threaten you. Maybe he could destroy your mom, but he couldn't destroy your mom's love for you. That was all you cared about, Phillip. Your mom's love. You don't care what happens to other people. You just care about maintaining those private temperatures inside yourself."

My hands were shaking as I handed the tool back to Rodney and he handed me another, like a surgeon and his faithful, highly qualified nurse in a long intricate operating theater while young students observed from a high balcony. I felt Rodney's hand grasp my shoulder.

"You all right?"

"I'm all right." I took the next few tools without examining them. I felt hot and dizzy. Magnified and eccentric, the motes swirled around me in the dim light of the television. The television volume was turned down low, and someone was whispering something about Islamic Fundamentalism. I applied the tools, one at a time, to Dad's flushed, warm skin. Gently, at first. My skull throbbed with a low dull ache which seemed to intensify with every move I made. It was a sharp, shooting pain at times, into my sinuses and eyeballs.

"You know what you did, Phillip," Pedro was saying, without bitterness and without remorse. "You did it, and now you have to know you did it. You have to know you did it, Phillip. Otherwise it doesn't mean anything. Otherwise it's like it never happened. Then I'd hate you, Phillip. Then I'd never forgive you."

My knees buckled slightly as the blood rushed to my feverish head. Pain expanded in my skull like the skin of a balloon. I handed the tool back to Rodney. I touched Dad's hot flushed skin with my trembling fingers. There was a large blue vein under his neck.

"He's still breathing," I said. "He's still got a pulse."

"Of course he does. You haven't *done* anything to him yet."

"I haven't?" The room was turning slowly around me. "I haven't done *anything?*"

I heard the tool clatter into the steel box. Then Rodney said disgustedly, "I thought you said you knew how to do this."

"He does," Pedro said.

"I do," I said, stepping away from the ambient warmth of Dad's trussed body. "I do, I really do." I was gesturing with both my hands, trying to make the room stop moving. I heard something kick. Weight was pouring from the mouth of the Budweiser can into the thick pile carpet at my feet. "I just need a drink, that's all. I think I'm coming down with something." I brushed my forehead with the back of my hand. "I think I may be coming down with a fever. I may be coming down with a cold or something." Tears were forming in my eyes, and I wiped them against my shoulder. I was backing out of the living room. I didn't look at Pedro as I passed him. Suddenly I was in the bright latex kitchen. All the lights were on. One of us had left the refrigerator door open, and the engine was humming, generating its icy mist. One of the eggs in the egg rack was cracked and exuded a yellow, inflating gel.

"Hey, Phillip! We gonna get this over with or what?"

I wanted Rodney to go away, but I knew he wouldn't, he wouldn't go away. I grasped the open door of the refrigerator and braced myself against it. The cold air rushed over me, like water in a bath. I needed this. I reached for a liter bottle of 7UP and took it to the beige-tiled kitchen counter. As I was reaching a glass down from the cupboards, I saw Dad's note on the counter beside the matching flour, rice and sugar bins. An empty pharmaceutical vial rested atop Dad's letter like a paperweight. The letter was printed on Dad's Epson printer, like junk mail advertising some new MasterCard Card Bonus Club Service. The empty vial said, "Phenobarbitone 50 mg. Take one in evening to sleep." Inside the vial was a thin white powder, like the powder you find at the bottom of an extinguished loaf of Wonderbread. I picked up Dad's letter. Dear Son, it said,

> *I hope this is something you'll understand better than your mother, who has a lot of other things on her mind right now. I just think that perhaps things will be much better for both of you when I'm gone. Please do not feel guilty about this even a little bit, since it is a decision I have made without you knowing it or having anything to do with it. I think this is the only way to provide a quick resolution for everybody, since I am certain I am*

suffering from some irreversible stomach cancer or maybe even something worse, since I can't sleep at all and my stomach feels terrible all the time and there are worse symptoms I won't really go into right now. Just remember that whatever happens wherever in your life, that your parents really did the very best they could to make you happy, it's just sometimes they couldn't stop themselves from being selfish, stupid or confused. All parents fail their children and we all have to get used to that, I guess. You'll have children of your own someday and maybe then you'll understand. I'm counting on you to take good care of your mother after I'm gone. I know you can do it since you did very well without me before I came to make both your lives so miserable. Everything is in very good shape with my lawyer, whose business card you will find attached, and whom I have carefully informed as to your mother's condition so he will keep a good eye on her from now on. There is $5,000 cash in my wallet.

Love,

Dad

I put the letter down on the counter. Immediately it began absorbing a semicircle of 7UP, causing many of the words to expand into nervy blue blotches. My hands were trembling, and I felt a slow descending warmth in my stomach.

I felt insecure and dizzy. Rationality had abandoned me like a boat or a train. I wanted to grasp hold of something, but I couldn't find anything firm enough. I heard the letter crumbling in my fist, the thin computer paper like something you'd wrap around steaks at the butcher's. I was hot with sudden steaming rage. Dad thought he was going to leave me. He was going to abandon me and Mom and the baby. He didn't care what happened to us, or what sort of place we might end up. He wouldn't even take us with him. Suddenly I recalled a solitary moment from my childhood. I was standing on the front lawn, watching Dad's car pull out of the driveway. I was holding something in my hand— a toy truck, or a plastic soldier, or perhaps a partially macerated baseball card. It was an offering, but he wouldn't take it. He was climbing into his big automobile and slamming shut the door. He was pulling out of the driveway, looking over his other shoulder, not even seeing me. Then I was watching the taillights of his car fading in the gray dusk. I called his name but he didn't turn around. I started to run after him. The streets were boundless, punctuated by simple trees and hedges. His car was pulling further away, he didn't see me. I was running further into the darkness. I didn't know where I was, or where Dad was. From that day forward, Dad stopped being Dad altogether. From that day forward, Dad became somebody else entirely.

"Hey, Phillip! This is your party, guy. We're still waiting in here. Bring me a little more Jack Daniel's—and while you're at it see if you've got anything munchable, you know? Pretzels, or sardines or something."

I returned to the living room with Ry Krisp, beer nuts, Hershey's chocolate kisses and a renewed sense of purpose. Dad thought he could go places without me. Dad thought he was a man and I was just a boy. Dad thought he was special, and I was nothing. I reached into the toolbox and took something which looked interesting. Then I stood again over Dad's unconscious and fleeting body, like a surgeon conducting life's most sacred rites.

"There you go," Rodney said. He was flicking beer nuts into his mouth with one hand; his left arm embraced his can of 7UP, into which he had poured prodigious whiskey. "Now we're getting somewhere. Ease up—we're in no hurry, right?"

Rodney's hand took my arm firmly. Just as firmly, I shook it away.

"Be careful, there. Hold on, you want me to go get some towels or something? Phillip? Are you even listening to me or what?"

I wanted to be good. Feverishly, as I worked, I knew I wanted to be good and to enter the kingdom of Heaven. "I used to pray," I said out loud, to nobody in particular. "I remember when I was little, I used to pray every night."

"What's that?" Pouring more whiskey into his soda can, Rodney dripped some onto Dad's white shirt sleeve. Then, after a moment, "Are you sure you don't want me to hold something for you?"

Again I shrugged him away. "I used to pray on my hands and knees, and imagined a Heaven filled with white lacy clouds. Many pleasant men and women came out to greet

230

me as I entered through these tall, alabaster white gates. There was a young girl there about my own age. I thought she was really beautiful, and we became close friends. One night she let me kiss her, and another night I saved her from the hordes of Satan's evil minions. I imagined all this while I prayed. I prayed to be good and pure. I wanted to remain a child forever."

"I never prayed to be good," Rodney said, reflectively sipping. "I only prayed for three things in my entire life. Money. Women. And power. And when you get right down to it, I'm not in such a hurry about the women and the money. Power's the main thing, Phillip. Power's the only thing worth really praying for."

"I wanted to do good deeds. I wanted to help cripples and old women who nobody loved. I wanted to save puppies from the pound, and teach broken birds how to fly and be free again."

"What are you doing with that—"

"I wanted there to be absolutely no pain and suffering in the entire world. Sometimes I wonder why I wanted that. I can't understand the dreams I dreamed then. What did they matter? What did pain and suffering have to do with me? They had nothing to do with me. They were things in the world, they were things different from me. I'm not really in the world at all, Rodney, am I? I'm really not, am I?" I was hot with dizziness and my own blood. With the back of my hand I wiped the sweat gathering on my forehead. I needed to lie down for a minute. I needed a glass of ice-cold water. But I couldn't relax just yet. I wasn't finished.

And then I heard the sudden crack of Mom's overpainted door opening down the hall. It all seemed perfectly natural—the world right now, events and circumstances. Then Mom's slow, balancing footsteps, her large stomach preceding her into the mouth of the living room. Rodney nudged me sharply with his elbow.

"Hello, Mrs. Davis," Rodney said.

I was just about to attach a pair of snub-nosed pliers. I cupped them in my hand like a guilty, smoldering cigarette and turned. Mom was standing there, watching us, her face glowing, wearing the big blue robe I had bought her for Mother's Day.

Mom was looking at Dad's face as if it were the face of a child in a photograph. Her eyes steadfastly refused to look at any of the things I had begun doing to him.

"When you're finished in here, baby, I want you to leave," Mom said. The fingertips of one hand were poised upon her stomach, as if it were gauging delicate reactions down there, secret chords and melodies, Morses of blood and plasma, protein and bone. "I don't want you around the baby. You can take the car, and all your father's money. But go far away, and I'll say I don't know what happened. I was asleep in my room. I woke up in the morning and found him. I hadn't heard a thing all night. I'll still love you, Phillip, but I don't want you around anymore. I've tried hard to understand, but I'm afraid I just can't understand anymore."

It was cold static moonlight. Outside, in the distance, I heard the helicopter beating past again. Somewhere in the night a police car radio sounded, and then its brief momentary eruption of inhuman voices: *"Not in the alley, over"*; *"Roger, Sam-six."* I was thinking, Just fine. You go where you want to go, Mom. I'll go where I want to go. I was staring her in the eye. I had nothing to be ashamed of. She, on the other hand, wasn't looking at me. She couldn't look at me, because she knew it too. She knew that this was my night, the night of my stark ascension. Dad wasn't going anywhere without me. Mom was the one who would be left behind. Everything was going to turn out exactly the way I wanted it to turn out, and there was nothing Mom, or Dad, or anybody else could do about it. I was going to have my way, simply because I finally understood in which direction my way led.

"I knew I lost you in San Luis," Mom said. She was watching the can of 7UP in Rodney's hand. "You'll never change, Phillip. You are the way you are, and that's that, I guess. I didn't say anything but I knew, and you knew I knew. Whatever your father knew isn't any of my business. Your father simply shouldn't have gotten involved. He knew better. He knew me. He couldn't have been that naive about you. I don't know what he expected when he came here, but we never invited him, we never made him any promises. Now, if you and your friend don't mind, I'm going back to bed. I'm going to sleep for about a hundred years. When you leave, remember to lock up. Don't leave any lights on. You won't be able to write me, because I won't let you know where I've gone."

Mom turned, paused with one hand on the wall, the

other on her stomach. Then, cautiously, she conducted my unborn sibling down the hall to her warm and silent bed.

There wasn't any time for reflection. I attached the snub-nosed pliers. Nobody was going to tell me what to do anymore. Nobody could send me away or leave me. Not even Mom. Not even Dad.

"Jesus," Rodney said. His hands were trembling, his voice faint, his eyes intent on my work now with either concealed admiration or blank distrust. "Your family's too much, guy."

"I always wanted to be good," I said as I feverishly worked, feeling vast geologic plates and fissures expanding in the earth deep under my feet. "I always wanted to go to Heaven. Now I don't care. I'll go anywhere. It's quite a relief, you know. It's like having all your appointments canceled and knowing you can spend the whole day in bed with a good book."

"You're really something, Phillip," Rodney said. "You really are."

"In order to free the self, one must abandon all preconceptions about what the self is." As I worked, the words arose in me without my volition. They were like the hard intricate tools I wielded, they were like the dense yielding body of Dad. Associative, crystalline, buzzing, hard. Next to these words, the world itself seemed to reliquefy itself, dissolving in the blood of some archetypal Christ. "Make no mistake about it—the self exists, Rodney, and this is it. *This* is the self. This is the self here." I showed Rodney something on the end of one stubby screwdriver. "Blood, tissue, bone, cartilage, marrow, mass, gravity, liquid, sound, light. It moves or it doesn't move. It lives or it doesn't live. This is the history of luminous motion, Rodney. This is the flux and convection of sudden light. We're all the same but we're all not the same too. What you know is not what I know. What you prefer is not what I prefer. There's just this—and this—and this"—Dad's body gave a sudden, galvanic kick—"or *this*," I said, enraged by the still pulsing life in him, "or this or this or this or this. This here, or this *here*. *This* is all we are, this warm and fragile envelope, this thin impacted tissue. It's not that we exist but that we know we exist that makes our lives so miserable. And this—this is nothing. And this, and this, and this. This is all nothing too."

"And this," Pedro echoed. "And this, and this, and this."

"This is the progress men and women make alone in the world of light," I chanted, the words filling me with heat and rage. "This is all we are, Rodney. This is all we'll ever be . . ." They were my words but they were somebody else's words too. Mom couldn't leave me. Only I could leave Mom. I was dizzy with fierce excitement. The blood coursed and raced in my head. I was moving too, through these humming veins, down these moist undulating corridors. I was moving into the world of Dad's body, a place even Dad had never been before. I would show them. I was going to show all of them. Mom and Dad, Rodney and Beatrice, Ethel and the world. The whole world, the whole vast and intricate world. And Pedro. Pedro, of course . . . I was going to show all of them. "This is it, Rodney. This is the light. This is Dad's light—but now it's mine. Now it's my light. Now it's Mom's light too . . ."

Rodney said, "Hey, Phillip. What's that noise?"

"This is the history of motion, Rodney. The history of motion. Look—the history of motion. The history, the history of motion . . ." I could feel it now. I knew it was coming. I could feel the pulse of it in my bones and skin. It raced in my blood. It raced in Dad's blood too. Something spurted into my eye and I wiped it away with the back of my wrist.

"Hey, Phillip. Something's wrong, man. Phillip. Hey— get it together, guy. I think somebody's outside—"

It was mine and it was Dad's, and someday it would be the baby's too. Me and Mom and the baby and Dad. And the light, the light—

and then suddenly there was just the awful thundering noise of it, descending outside in the ruined and brilliant white sky. Finally I saw it. The light, the hard bright white light flashing through the cracked venetian blinds into our living room, beating and flashing, fast and sudden and secure, roaring and louder like massive engines driving and obliterating everything, even the night. It was all life, it was all living. This was my life. I was doing it now. I was living my own life now—

"Jesus Christ!" Rodney shouted, grabbing my hands, pulling at me. "Phillip—we gotta get out of here!"

But nobody pushed me, nobody made me do anything I didn't want to do. Not tonight and not ever again.

236

"Rodney!" I shouted. I was disentangling my hands from Dad's clinging warmth. I even tried to lift Dad in my arms to show him. I wanted Dad to see too. I wanted everyone to see. "Look there! At the window!"

It burned through the window blinds. It was life. It was white. It was coming for me, for me.

"Rodney! Look!"

I was shouting over the noise of the beating helicopter rotors. Everything was so simple now. All the hard eternal light of it was burning in our blood and our bones and our brains . . .

"Rodney!" I called. I turned around the room, alone in the whirling hallway. The back door to the garage stood wide open.

The light thundered through everything, beating back drapes and curtains. The entire house was rattling and trembling, the light swirling and turning. Pedro, darkness, Pedro, darkness. And then just the darkness. And then just the light. And Dad's life hot on my hands and my clothes and my face, and the hard beating light outside like a summons, celestial and vast, like Jesus or God, Buddha or Muhammad. Like Dad's voice. Like Mom's love. Like light and motion, motion and light—

"Rodney!" I cried. "Come back! This is it! It's over! They're here!"

"ROGER, TEN-FOUR" the shortwave outside blared, its cessation as suddenly loud as the world around it. Massive car doors opened and slammed, footsteps sounded heavy and fast on the front stairs, I heard a flowerpot crash into

the cold alleyway. "Open up!" they shouted. "Open up in there!" And then that loud peremptory knock at my front door and, as I turned, the door exploding open with a crash of dark large-bodied men in dark blue uniforms. Their badges and flashlights gleamed, their weapons flashed, but I wasn't afraid, because I wasn't going anywhere I didn't want to go. They were mine—I wasn't theirs. My arms were even outstretched to embrace them, my hands and face stained with the sacred blood. *Totem, totem, totem*, the hard beating wings resounded outside our fragile home. *Totem, totem, totem* . . . and the light streaming through me like the rain and the wind and the sky. We would all taste the Eucharist, we would all ingest the flesh and suffer strange transubstantiations. We would all find God, we would all live forever. I knew she would tell them. I knew all along she'd never leave me or send me away. My mom loved me. They were here. I was saved. It was the police.

THE
HARD SONG

26

While undergoing three weeks of isolated observation at Valley Youth Correctional Facilities I was allowed some books, a pen and notepad, and a few choice hours each afternoon of strictly regulated media privileges. I was also granted, almost as an official afterthought, what seemed to me at the time like virtually acres of soft, casual introspection. They tell me I slept nearly two full days and nights upon my arrival, awaking only to take slow bites at the facility's tepid, customized meals. I don't, however, remember those first two days at all. I only remember waking one bright spring morning to the harsh sun flashing outside my window, the glass of which was inlaid with a fine protective wire mesh. The thin bed and walls of my room seemed drab by comparison. I heard a few singing birds, eccentric, anxious and shrill. I was confronted by a long mirror in which I sat on my bed, observing myself with a sort of cool diffidence, as if I were warden to my own reflection. I assumed that invisible behind the mirror sat my more official audience. My reflected face was lined and bruised from excessive sleep, and I poured a glass of water from the blue plastic pitcher beside me on the weak, clumsy bureau. My room was like the rooms of the motels in which I had been raised, at once transient and profane, fleeting and ill-designed. After months of strange vacation I was finally home again. No matter how much fun you have on vacation, it's always good to be home again.

"Do you know what you did?" Officer Henrietta asked me soon after my arrival. My afternoon sessions with Officer Henrietta, a trained and certified psychotherapist, became the only certain ritual of my day beyond mere self-maintenance. Officer Henrietta was a bluff, affable man, but one who wanted it known he wasn't about to take any nonsense, and especially not from a child.

"I don't remember," I confessed. "But I'm sure that, whatever it was, it was wrong. Or else I wouldn't be here, would I, Officer Henrietta?"

"What you did was very, very wrong," Officer Henrietta said. "What you did endangered the lives of people you loved. What you did frightened a lot of people. It frightened you, so you can't even remember what happened. Do you believe you're capable of that? Do you believe you're capable of doing things so horrible you can frighten yourself that bad?"

The office in which we met six days each week was cluttered with papers, Styrofoam coffee cups and crumpled Hershey's wrappers. I always felt comfortable in that office, and

242

actually looked forward to the rather easy, meaningless conversations Officer Henrietta was kind enough to conduct with me. Officer Henrietta's distinct, often provocative questions never startled me or made me feel ill at ease. Instead they always implied what my own obvious responses simply had to be, responses I did not utter so much as activate, like functions in a computer. Graphs, data, production, profit, loss. Anger, love, resentment, sadness, pain. The world of the self and the world of machines. During these days and nights of slow, unhurried reflection, I began to realize that those were the two worlds I always seemed to be getting confused. The world of the self and the world of machines.

"I believe the human mind is capable of anything," I told Officer Henrietta. "The mind is its own place, just like Milton's heaven. It sounds like something Blake would have said, doesn't it? I'm talking William Blake, now. Do you know who William Blake is, Officer Henrietta?" Officer Henrietta's black felt pen sat poised at the edge of the paper, but his unresponsive brown eyes were trained upon me with a remote, unfocused expression, as if they were staring into something both vital and abstract, like the weather. "And as for me," I said, "I believe I'm capable of anything I'm capable of doing. I can be anybody I want to be, because only I have the power to decide. Not the world, not this institution, not you and your framed documents. Only me and my conceptions of me. My mom taught me that. I can grow up to be a doctor, an astronaut, or even the president of the United States. I can be a bird, a rock, a cloud. I can be anything I want to be, Officer Henrietta. And I'm afraid there's not really anything you can do to stop me."

Every once in a while Officer Henrietta emitted long, mystical sighs, vague punctuations which indicated vaster and cooler worlds than ours filled with sunny, padded white clouds and sparkling blue beaches. He leisurely chain-smoked Marlboros or Winstons and drank vile, bitter coffee just dimly discolored by Cremora. He showed me ink blots and asked me what they meant (though I suspect he may have known already without my help). I described for Officer Henrietta bats, abattoirs, leering faces and dark twisted passages filled with incessant and secret motion. He asked me abstract questions. If you drew a picture of yourself on a piece of paper, what color paper would you choose? If a strange man came up to you on the street and asked you to love him, what would you say? If you were on a sinking ship, who would you save first—the women or the children? These were all fine questions, and I answered them the best I could. I told him I would choose a sheet of beige paper, because that was the color of my mom's car. I told him I would tell the strange man to love himself, and let me get on with my own life. I told him in the event of a shipwreck I wouldn't try to save anybody, I would let the whole world drown. We would all return to the deep earth together, drifting down through the intricate seaweed and glistening blue water, women and children all together at last, journeying into a safer and warmer world than the one of broken ships.

One day Officer Henrietta began showing me photographs of a beautiful woman smiling with white, white teeth. He showed me photographs of a man tied up on a nice sofa in the living room of a nice house. Strange things had been done to his body, from which the clothing had been torn in places, like the paper windows in a Christmas Advent calendar. The pictures seemed slightly familiar to me in a dozy, unimperative way. I thought vaguely I might like to meet these people. But then I also thought it wouldn't matter to me that much if I never met them at all.

"Is there anyone you'd like to see?" Officer Henrietta asked, laying the photos facedown on his desk, shuffling and stacking them meaninglessly like a deck of cards. "Can you think of anybody offhand? A relative, maybe. Somebody you especially love."

I thought. I thought about silent places where darkness covered everything with an oily film. These were uninhabited and soundless places, like hidden chambers of the moon. You could see no faces there, you could hear no names. Like Officer Henrietta's photographs, these places didn't matter to me that much one way or the other.

Officer Henrietta was looking at me. I looked at the sheet of unmarked white paper on the desk before him, at his upraised and ineffective pen, at a photograph of his beaming family in a cheap plastic frame.

"Is Rodney around?" I asked him after a while. "Is there any possibility I might be able to see my friend Rodney for a few minutes or so? I'd like to know how's he doing, you know. I'd just feel a lot better if I knew my friend Rodney was all right."

Alone in my discrete room I would lie on the bed for hours gazing at the stale ceiling and talking to the figures who sat observing me behind the mirror. I could hear efficient machines whirring back there, official documents being filed into sliding cabinet drawers, the occasional hum of a word processor, the brief interjections of a clattering typewriter. "I think I'm learning to take things a lot easier than I used to," I told these invisible people. "In the past, I may have been too quick to make judgments. I couldn't seem to feel satisfied to accept the way things were. I think I've learned a lot about myself in the past few days or so, and I may be on the verge of some real sustained growth—both intellectually and emotionally. I'm growing more and more interested in Eastern religions, for example. Yoga, Brāhmanism, Buddhism, Tantrism, Oriental alchemy, mystical erotism. We're very much a thing-oriented culture—the West in general, I mean. We're into making things, changing things, moving things from one place to another. Sometimes I think it's best just to let everything lie. To not keep banging and bumping away at the world, to accept things for what they are. I guess in a way that might sound sort of escapist to you. I'm sure my friend Beatrice would probably be quick to agree. To imagine the world and all its suffering as a sort of necessary trial, one which presumably conditions us to understand our true *being*, is to imagine that the world itself doesn't matter, nor the conditions in it. That means, in a way, accepting the world's cruelty and its pain. That means just leaving it alone to get on with its own alien and material processes, however wrong and unjust they may be. I'm sure that could sound rather self-centered, even pretty ambivalent or smug. But I think there often comes a time in your life when you stop worrying about whether the way you think is right or proper or not. You just get tired, and start accepting the way of thinking that's easiest and least worrisome. Maya, the world as illusion. Karma, that duplicity and evanescence of mere physical life, the incessant beat and blur of material repetition. Then nirvana, the self's final liberation, a dream of nonbeing as pure being. We find our way out of this world within this world, I guess that's what it all boils down to. Now that I've got your attention, maybe I can ask you to send me a few books. I could use one or two on Vedantic philosophy. Then of course *The Upanishads*, and the *Bhagavad Gita*. I could use a general edition of Patañjali's *Yoga-sūtras*, while we're at it. I don't mean to hurry you, but whenever you get time. I was thinking they might even make a good permanent addition to the library here. I mean, when kids get screwed up like I do, I think they need to turn to some sort of traditional wisdom in order to work through their confusion. Kids who try to break the rules are only trying to find better rules they can live by, and I think the best rules are always the ones you carry within yourself. Kids need to learn they can't expect anything from anybody. They need to learn that everything they'll ever have is already inside them, is simply waiting there to be recognized."

As I lay in my bed talking, I could hear doors opening up and down the hallway outside. I could hear toilets being

247

flushed, and gurneys squeaking and clattering along the polished tile floors. The halls were filled with the bright, audible noise of institutional fluorescents, that hypnotic artificial sort of light you might discover like atmosphere in some alien space station.

"Of course, you know what Beatrice would say about all this, don't you?" I shrugged, affecting a smug disconcern. I couldn't remember if I'd told these invisible people anything about my friend Beatrice or not. "She'd say all my talk about some spiritual liberation's just a big con, that I'm trying to disavow ontology. You can't disavow ontology—that's what Beatrice would say. Ontology's what happens when you're hit by a bus. It's not something you can just disavow."

I wanted redemption in these days of my slow recuperation, the warm equatorial haze of samadhi, the total cessation of all transformations. They never brought me the books I requested, however. Whenever I reminded him, Officer Henrietta avoided the issue. Instead I received a few "world classics." *Les Misérables, David Copperfield, War and Peace,* all of them abridged and illustrated for some theoretical "young adult" reader. These books lay casually disregarded on my bureau while I lay on my bed thinking. If I could not learn redemption, I could at least imagine or even reinvent it. I gathered what fragments of Oriental wisdom I could recall and tried to generate larger worlds around them, vaster pictures into which these fragments might tidily fit,

248

like pieces in a jigsaw puzzle. The mind was just a reaction of pure spiritual being to the world's material force. The mind was a whirlpool, constant and uncontainable, which spun off into the world knocking into other things, inciting other spirits to move. This was karma then, I decided: the constant push of objects which tried to make of spirit an object too—a sort of cosmic bullying, a rushing and herding of things into other things. These forces made life, death, people, pain, suffering, cities and, worst of all, human emotions. They made anger. They made hate. For years now I had been filled with this hot and irreproachable anger that burned and flared in me without warning—this anger I could not contain, which had caused me to do something, or perhaps even a series of things, for which I had been legally incarcerated. I had been incarcerated in order to protect the people I loved, and as a result of this real burning drive in me, this raging drive to hurt, to conquer, to create a more material and corrupted world, I had harmed people—people I loved, Officer Henrietta liked to remind me—and I had consequently harmed myself as well. You can't direct your hate at other people; hate is a force that burns him who uses it too. Hate never does anyone any good, I thought. This was the lesson I had been brought here to learn, and I was amazed at the effortless and benign nature of its composition. It had simply grown in me, blossomed like flowers. It actually took root and grew from the very rage and anger it was intent on eliminating. Everything carries within it the fuel of its own driving antithesis, I thought. Anger is the stuff from which real love and knowledge grow. In order to grow and learn, we must permit the world to betray itself.

On a Friday morning before lunch I was permitted two visitors. Beatrice and Ethel, both immaculately buffed and manicured, lipsticked and glossed, sat on cracked vinyl chairs in a small Visitors' Lounge which included a wobbly, unvarnished pasteboard coffee table, some magazines and an additional cracked chair for me. Ethel wore a gauzy hat and was potent with cheap perfume. Beatrice wore a dress which at first appeared very bright on her, and then, after a few minutes, as I grew inured to this uncustomary sight, began to appear slightly cheap and shiny, like polyester or cheap lacquer. Her hair was washed, her lipstick excessively bright. As I sat down in the empty chair, we all regarded one another uneasily, like strangers brought formally together by some parent-teacher committee or charity bazaar. Ethel cleared her throat and I examined my pale hands in my lap. Beatrice could not remain silent very long. She shifted nervously in her seat, adjusting and readjusting the hem of her awkward skirt.

"I don't think I owe you any apologies, Phillip," she said. "Don't think I came here to apologize."

I wanted to tell her about my anger, how it had departed suddenly, become a force of mere matter. I, meanwhile, was growing more ethereal and abstract. I thought she might like me better now, this "new" Phillip.

"That's OK, Beatrice," I said. "I didn't expect you to."

"Then you expected right." Beatrice's blue eyes flashed

at me. "What I did I did for your own good. I don't give a damn about your father. You could have drawn and quartered that SOB for all I care—"

"*Betty*," Ethel cautioned her abruptly.

"—and I really *mean* that," Beatrice continued, cautioning Ethel back. "What I *do* care about is you and Rodney. I was damned if I was going to watch you both throw your lives away over your stupid father. He wasn't worth it, Phillip. You're trying to kill the only person in the world you love because the world won't love you. You're a patent narcissist, that's what *you* are. You gaze at the world and expect the world to gaze dreamily right back at you. You've got to grow up, Phillip. You've got to learn to relax. You've got to start showing some real concern for people in the world who weren't born with all the advantages you've had. Think of the children in Soweto and Afghanistan. Think of the political prisoners throughout Latin America and Eastern Europe. The world's not reflective, Phillip. It's dynamic and blind and stupid and correctable and utterly forlorn, just like you. Just like me and just like Rodney and just like Ethel here—" Ethel blushed slightly, as if she were flattered simply to hear her name mentioned in any context at all. "It's a world with real problems, that causes real pain, that promises real pleasure and abundance. I haven't been able to sleep all week. I knew you knew, even without me telling. I just hope you know too that it's not because I don't love you. I love you and Rodney very much. You're my family, and if I had to, I'd call the cops again today, right

251

this very minute, if I thought you and Rodney were about to do something you'd both regret later. You can bet on it. If you tried to pull the same crap all over again, I'd have the cops all over you in a second. I'd see to it they were all over you like a cheap suit."

Meanwhile Ethel snuffled behind a dingy Kleenex. Her eyes had grown moist, and her mascara was starting to run in places.

"Where's Rodney?" I asked. It wasn't as if I were addressing her with my question, but rather trying to push her out of my way. "When am I supposed to see him, anyway?"

The Kleenex in Ethel's hand began violently shaking. Obviously flustered, Ethel looked from Beatrice to me, and then at Beatrice again. Her face was very pale.

"They caught him climbing over the back fence into a neighbor's yard," Beatrice said. "He's in a holding cell here, just like yours. But early next week he'll be transferred to a separate facility altogether. They don't want you and Rodney seeing each other again for quite a while, Phillip."

"It's my fault," Ethel said. Finally she was looking at me. Her damp hand touched mine. "I shouldn't have let him near you. I know you don't understand what you do to people, Phillip. But you were a terrible influence on Rodney. I should have paid more attention to you, Phillip. I should have gotten to know you better."

"When you get out, Phillip, I want you to call me." Beatrice leaned forward earnestly in her chair. It was the customary intensity of her expression now which made her stiff dress seem more and more like a disguise. "I want you to call me and tell me where you are. I left word with your parents, but they're not responding. When you get out, Phillip, I want to see you. Call me, promise? Call me first thing."

I promised I would. I went to the door and asked the guard to take me back to my room. But already I knew I would never call her. We were all moving off into separate worlds and galaxies now. We were all journeying off to find the only redemption any of us could afford. I didn't owe Beatrice anything and she didn't owe me. And I certainly never wanted to be betrayed by Beatrice ever again.

A few nights later, while I was lying on my bed trying to design my own mantra, Rodney came and tapped at my door.

"Phillip," he whispered. "Hey, Phillip. It's me."

My room was dark, and a mantra to me was just a faultily remembered notion. A mantra was a puzzle which drew your mind into the deeper, more complicated puzzle of the world itself. Or so I understood it. You are hanging from a rope in a dark pit. At the bottom of the pit a lion roars at you. At the top, a mouse gnaws silently at your rope. This is just the universe you've landed in. Even salvation doesn't last very long.

"Hey, Phillip. I've just got a minute. Are you up?"

I climbed out of bed and went to the door. I could sense

the pressure of Rodney's fingertips on the outer side of my hard steel doorknob. In my dark room with the door bolted from the outside, I thought of Dad.

"Rodney, I don't think I'm supposed to talk to you."

"Look, I only got a minute. I gave the guard twenty bucks from the cash Ethel smuggled in for me. Here, something else."

I felt Rodney's body brush lightly against the outer surface of my door. Then I heard something strike the air vent over my head. Rodney's feet landed softly on the outer hall. Then I heard Rodney jump again, and a pack of Winstons popped through the air vent and ricocheted off the dark, wide-screen mirror.

"You get them?"

"I think so."

"Just one thing, Phillip—and listen to me real carefully now. Tell these fuckers anything they want to hear. You got me? Anything. Be anybody they want you to be, and get the fuck out of here. This place is full of crap, man. Have you gotten a look at that Officer Henrietta, Boy's Best Friend? I never met anybody filled with more crap in my entire life. So do what they say, right? Say whatever they want to hear."

"Rodney—I have to ask you something—"

"Hold on a second—" Rodney whispered. Then his voice grew dim, harder and echoing as he leaned away from the door. "What?" he asked, in a louder voice. "Just one minute."

"Did we hurt anybody, Rodney?" I asked quickly, afraid

my question might go unanswered. "I'm beginning to suspect we may have done something we shouldn't have—"

"*Fuck*, lady!" Rodney told the echoing hallway. "I told
you *one minute!*"

Then, as if the institution itself was hurrying to divide
Rodney from my vital inquiry, I heard the dark weighted
body of a guard sweep down the hall and Rodney's elbows
knocking loudly against my hollow door.

"*Fuck* you, lady. You fucking *cunt*, I *said* I'm coming.
Watch it there, will you? *Shit!*"

"Rodney, keep in touch," I said, but he didn't hear me.
His voice was steadily diminishing down the long corridor,
as if someone were slowly turning down the volume on a
radio.

"Tell the fuckers anything, Phillip! You hear me? Tell
them whatever the fuck they want to hear and get yourself
out of this dump. *Hey*, lady. That's my *arm* you've got there.
Know what you just did, lady? You just *screwed* yourself out
of twenty fucking bucks—that's what you just did. You blew
it, lady, because I'm going to *eat* it, you hear me. I'm going
to *eat* your twenty fucking dollars before I'll let you see a
piece of it."

"I'll do my best," I whispered secretly, as if Rodney and
I were conferring in one of the unoccupied, transgressed
homes of our childhood. "I'll tell them anything—as soon
as I figure out what it is they want to hear."

I would call Rodney's house months later after my re-
lease, but his number was long disconnected by then. My
letters were returned by the post office, no forwarding ad-

dress available. I didn't blame Ethel, really. I was not the only man in her life.

Then, just before it disappeared forever from my world, Rodney's voice said, "Hey, lady. I thought you were supposed to be letting me out to take a piss."

27

The following Thursday I was escorted to the Youth Facility's remotest, somberest corridor for what Officer Henrietta called an "informal prelim." Escorted by a young woman in a patently unattractive and bulky blue uniform, I saw for the first time the general design of the institution, filled with its atmosphere of harsh fluorescent light, bracketed by white ungleaming tile floors and lacquered beige walls. In one corridor I heard the monotonous, aggressive clocking of a Ping-Pong ball, and in another a sports program on TV. We passed a cafeteria where young Chicano men wore white aprons and black hair nets. They had tattoos on their muscular arms and lean, mustached faces, and swabbed down the floors with tall wet mops. They rested the mops against their shoulders and looked at me as I passed. I felt very uneasy looking back. They scratched under their arms and continued watching me go by. Then they scratched roughly between their legs.

In the distant and official temperature of the Deposition Room (or so the plastic sign said on its door) I was seated in a stiff chair at a table with three men who wore modest navy-blue suits with modestly patterned ties. They were introduced as a judge, a prosecuting attorney, and a Youth Offense Adviser, who, I assumed, was sort of like my lawyer. When they spoke, however, I could never distinguish who was saying what, or from what official position.

"There's no prior record, then."

"None."

"If the parents aren't making any charges, what are we holding him for?"

"We found illegal substances on his friend. Everybody thought it might be a good idea to keep him under observation for a while."

"Who's everybody?"

"Joe, Harry, Delacruz. Me."

"That's not everybody, is it?"

"No, I guess not."

"How old is he?"

"Eight years old."

I sat a little forward in my chair. "Almost eight and a half," I said, but they didn't seem to hear.

"What does the psychological profile say?"

One man opened a manila folder and read to us. "Severe amnesiac reaction to stress and poor body management. Perhaps a paranoid schizophrenic, with delusions of grandeur and competitive reality disorder." The man closed the file. "The doctor thinks the kid could grow out of it."

One of the men tapped his pencil against the flimsy Formica table.

"If we keep him, where are we going to keep him?"

"Here, I guess."

"What about a foster home?"

"I don't think this case is exactly cut out for your usual foster home."

"What do the parents think?"

"The father wants to take him to his home in Bel Air. He's pretty well off. He says he'll bring in hired help. He promises to keep a close eye on the kid for the next few years or so. He's going to hire private tutors and keep him out of school. He's going to bring in clinical psychologists. One doctor he's hired already wants to try some things on the kid with insulin. I say we give the kid back to his parents."

The men grew silent for a while. I sat quietly in my chair, trying not to look at any of them.

"What do you think, Phillip?" the most self-confident of the men asked me after a while. He crossed his hands on the thick manila folder. His hands formed a wedge which was aimed at my heart. "You've heard us discussing your future. Have you any ideas about where you'd like to spend the next few years of your life?"

I looked at the man. His dark hair was cut short in a bland, official style. He wore thick plastic-framed glasses.

"I guess I don't really," I said. "I guess I'll be happy to do whatever you gentlemen think best."

The warm reality of my small, institutional room was growing dimmer and more fluid each day while I lay on my bed, contemplating my freshly designed mantras and the world's annihilation and rebirth in the form of pure, rarefied and immaculate spirit. I was no longer concerned with what crimes I had committed, nor what penalties I might suffer for them. Life itself was a penalty of sorts, and the wide world its own infallible crime. We were all in it, we all lived it. Perhaps we weren't all responsible for this awful mess, but if we weren't responsible, then I thought it pretty safe to assume that responsibility in itself didn't really mean anything. Like Mom, we did not die or cease to exist so much as awaken to a more enduring and unfathomable life. Once we awoke, this world wouldn't matter anymore. We wouldn't even know where we were. We wouldn't remember who we had been, or care too much about what we were to become. Being would only matter then, and nothing else. Now that I was saved, now that Mom had sacrificed herself in exchange for my firm redemption, I believed salvation was possible for the world too. I believed anyone could find ultimate happiness, just like my mom had, just like I was finding it now.

"Tall," Officer Henrietta told me.

"Short," I said quickly. I liked this game because when I played it I could feel myself starting to disappear, leaving nothing but my automatic words behind. I could briefly glimpse the world into which Mom had vanished. If you played the game long enough, you completely forgot you were playing any game at all.

"Big."

"Little."

"Dog."

"Cold."

"Friend."

"Suffering."

"Hurt."

"House."

"Desire."

"Warmth."

"Father."

"Blood."

Officer Henrietta put down his stack of file cards and made a succinct notation on the plywood clipboard. He took his cigarette from the ashtray and looked at me.

"Mother," he said. He took a long drag from his cigarette. He did not exhale right away. He squinted a little, as if he were trying to peer inside me.

I looked at Officer Henrietta. I deeply desired one of his menthol cigarettes, but I had recently decided to give up nicotine, as well as all earthly substances which made me indebted to mere matter.

"The history of motion," I said after a while.

Officer Henrietta exhaled the smoke slowly. A fine, gritty mist expanded in the flat spaces between us and then evaporated. "What's that?" he asked. He was talking more slowly now. We had begun playing a different sort of game entirely.

I held my folded hands braced between my legs. This room was very cold all of a sudden. I suspected someone in

261

some deep, secluded and inviolate security nexus had acti-
vated the formidable air-conditioning.

Officer Henrietta and I continued to gaze at one another
for almost a full minute.

"Nothing," I said. I watched Officer Henrietta crush out
his Kool in the glass Denny's ashtray. "Can I please go back
to my room now?"

Officer Henrietta sighed. Times like this I felt very sorry
for Officer Henrietta.

One morning I was escorted not to Officer Henrietta's office
but rather to a wide asphalt parking lot where a gray, nondescript car awaited me. Without further ceremony or dis-
cussion I was transported to the busy corridors of a Valley
hospital, through avenues of pale sunlight and fading adobe
shopfronts and streets filled with cars and potholes and roaring buses. From the outside the hospital appeared very dark
and ominous, like some corporate office building with dark
cement walls and reflecting glass windows. I was taken into
a cool lobby and then into an elevator's tinkling Muzak.
Women and nurses smiled at me, then frowned at the stern
obdurate guard beside me who gripped my pale hand. I was
taken to a private room where a beautiful woman lay ex-
hausted in a bed, and a man and another woman stood at
the window beside the bed. When I entered the room,
everybody looked at me.

"I'll be right outside," the guard said. She was a woman.
She went outside and closed the door.

"How are you, Phillip?" the man said. The man wore a

262

large bandage across the right side of his face, and there were a few visible lines of stitches across the bridge of his nose and down one side of his neck. A second woman with sun-blond hair like the man's stood beside him, her arms folded, and glared menacingly at me. This was her room, she seemed to be telling me. These were her friends, this was her family. I wasn't really wanted here. At any moment, she could ask me to leave.

The man stepped forward, and I heard other bandages brushing dryly underneath his clothes. He stepped with a slight limp. He held his body stiffly. He took my hand and sat down with me on a pair of chairs beside the bed. "Your mother's still asleep," the man said. "She's all right, though. Don't be frightened. At three thirty this morning your mother gave birth to a nine-and-a-half-pound baby boy. Isn't that exciting, sport? I wish you could have been there. It's such a miracle, watching a baby being born."

The man sat there staring at me, but I just looked at the sleeping woman in the bed. She was very beautiful, for those who like women with dark hair and rather fair skin. Her hair was a mess, though. And without any makeup she probably looked a lot older than she really was.

"This is my sister," the man said, indicating the severe woman beside the window. "This is your Aunt Sally from Phoenix."

Aunt Sally hadn't taken her eyes off me yet. She was packing a cigarette against the ledge of the window, just sitting there and looking at me. Her cigarette went tap tap tap. A sign over the sleeping woman's bed said NO SMOKING

PLEASE. If Aunt Sally doesn't like the way I look, maybe *she's* the one who should leave, I thought.

Aunt Sally showed the man her cigarette. "I'm going outside for one of these," she said. It was as if she had read my mind.

The man nodded at her and she left.

"I'll be right outside," Aunt Sally added, and closed the door behind her.

The man and I sat together for a while and watched the sleeping woman. The man held my hand in his, and I didn't mind. I knew he was trying to tell me something. He was waiting for the right moment. He thought he might be able to detect that right moment in the pulse of my hand.

"We're all going home together, sport. It's a great house. It's in the best part of Bel Air. I'm sorry you've had to spend so much time in that awful place, but I didn't really have any choice. I've had a couple of good lawyers on it, and you should be able to go home with your mother and me tomorrow. I know things have been very confusing for you, sport. No hard feelings, I promise. But if we're going to sort things out, we're going to have to sort them out together, if you know what I mean. We're going to have to work through things together in our own house, just between us three—us *four* now, I should say—and not give up until we get it right. You follow me, sport? Are you with me on this?"

I didn't answer. I was already growing bored with looking at the sleeping woman, so instead I gazed outside the window at the white, cottony sunlight suffusing the San Fernando Valley. There was no color out there anywhere today, I thought. It was one of the Valley's white days.

"Would you like to go with me and see the baby?" the man asked. "We could do that right now, if you want. Before they take you back."

I didn't like this man very much. I knew that right away. I didn't really dislike him that much, either, though. I tried to be as tactful as possible.

"Let's just not rush things, OK?" I said. Out in the distance, I saw an oblique dark shape beginning to emerge from the white sky. After another moment I recognized it. It was the Goodyear blimp.

Two days later Officer Henrietta shook my hand and gave me a little lecture about growing into manhood and all the responsibilities a young man faces. Growing up is never easy, and young men face difficult problems every day, problems like sex and drugs and peer pressure. It was important to remember, Officer Henrietta explained, that a young man must learn to find truth within himself and his family, and that if a young man only knew that people around him really did love him, he'd also know that no matter how hard the problems or how difficult the choices, he would still find his way safely through any unpleasantness the world might have to offer. "Even when you're an adult," Officer Henrietta said, "it doesn't get any easier. You keep thinking it's going to get easier, but it doesn't. I think you just get used to the pressure after a while. I think you just become a better judge of your own character."

In the long pause that followed, I said, "I'm sure you're probably very right, Officer Henrietta. And I'll keep your good advice firmly in mind. I really will."

"Good boy, Phillip. Now you go along and pack your things. And if you ever have any problems, or if you ever have *any* questions and you don't know who to turn to, you can always call me. OK? Now get going. And keep in touch."

"I will, Officer Henrietta," I said, gazing for the last time around his blithe, cluttered office. I would miss it here. Here the games were always clearly games, and never really mattered that much. "And I will keep in touch. I promise you. I really will."

28

Beatrice was right, of course. I was going to grow up.

I moved to the big house in Bel Air with the man and woman from the hospital and was enrolled in a private school. I was given my own room, color television and VCR, and three times a week I was visited by a battery of psychologists and dieticians who examined me with very clinical smiles. I grew accustomed to the spaces and geometries that lay around my large house, and for the first few months I was occasionally allowed to explore these spaces, by foot or by bike. I was never alone, however, for wherever I went covert men and women followed me in slow, very obvious automobiles. Sometimes I even saw these men and women parked beside the schoolyard where I would sit during lunchtime recesses and watch my addled, utterly inefficient classmates run their races and enact their imaginary dramas of pirates, cowboys and tycoons. The covert men and women never bothered or oppressed me. I knew they were only there to protect me. Eventually they stopped coming around and I felt a certain calm emptiness surround me; I even missed those covert men and women in a way. It was as if I had lost the only authentic family I had ever known.

This was my home and this was my family where I did not really live so much as circulate among things, events and strangers like a sort of atmosphere. Here was the man in the chair by the fire. Here was the woman in the bed near the TV. Here was the baby in the room filled with bright plastic toys. The baby was very remarkable, and everybody always said so. It never cried or raised a fuss, and whenever you spoke to it, it seemed to know exactly what you were saying. "My name is Phillip," I would say some nights after everybody else had fallen asleep, standing alone over the baby's dark crib. "I live in the next room. Your parents support me and see to my education. When you grow up, you will be very happy. You will exercise and eat right, and be involved in all sorts of extramural sports at your school. You will fill this room of yours with many sports trophies and citations of scholastic excellence. You will eventually become involved with a pretty girl from your high school, and you will often bring her by the house. I will always live here, too, but I will never bother you. I will always be in the next room. I will always be a moment away, in case you ever need anything, or in case any sort of emergency arises. But otherwise, I think, it would be best if we didn't see each other too much. I'm not trying to be antisocial. I'm just considering all the complicated logistics involved in people living together over a long period of time." The small, intelligent baby would look up at me as

I talked. Its dark, attentive eyes concentrated on my moving lips. This was a baby the man and woman of my house would eventually be very proud of. They would never have to feel nervous about a child like this. They would always know where it was, and generally what it was thinking. They could engage in casual conversations with it, without worrying so much about what they said, or what it might say back.

I had a future now, as firm and incontrovertible as my house and my family. I would complete grammar school, junior high, high school. Perhaps I would attend USC or UCLA, and earn my degree in law, medicine or journalism. I would marry a lovely, patient woman who would bear me no more than three lovely children. I would acquire a good job, my own big house, and two cars in a two-car garage. A Pontiac and a Volvo. My wife and I would send the kids to summer camp every year, to give us a little time to be together. On Christmas, we would take everybody to the house of the man and woman who had raised me in Bel Air. We would drink and sing Christmas carols. Every other year or so either I or my wife would have an affair with someone, usually someone I worked with or my wife met at one of the various regional political and charity functions she often attended. We would consider calling everything off. The house, the marriage, the formal avowals. But then we would start to grow more anxious and uncertain the further and further we grew apart from one another. We would begin to feel ourselves verging on vast unlabeled places that seemed to open up out of the earth under our feet. We would come to tearful and sudden reconciliations, reconciliations that grew quickly more formal and sensible as succeeding weeks passed. Our children would grow up. Just like me, they would raise families of their own.

I have never been truly unhappy since I have settled down to a more normal childhood, but perhaps, at times, I do feel a little restless. On these nights, when my parents are asleep, I take out the little red sports car the man recently purchased, an MG convertible with a hard, racy little engine and quick catlike traction. I drive it out along the coast highway, or across Sunset into Hollywood, where the tacky streets are empty and somehow magical late at night after all the hookers and junkies have gone home, like the stage of an abandoned movie set. Some nights I drive south to Orange County on Highway 5, or even as far as San Clemente. The air is always pleasant at night that far south, clear and warm. There are still a few rolling hills and green pastures that have not been converted to barnaclelike condominiums, shopping plazas, hotels or bowling alleys. It is always nice just to drive and relax and not feel in a hurry to be going anywhere. It's nice just to drive aimlessly around for a while with my own abstract thoughts and dim, fading memories of a life that has always seemed to me rather formless and abstract to begin with.

Some nights, though, I drive to the San Fernando Valley and the house where Mom and I once lived together. The front yard has been reseeded, and a number of pine and fig saplings have been planted around the front yard, where they have already grown into substantial trees. The basement window-latch can still be slipped open with a flat screwdriver and, inside, the garage has recently been swabbed out with solvent. Cleansed of its familiar smells, even the familiar angles and architecture of that garage seem strange and unfamiliar to me now. A large Ford Galaxy automobile stands in the middle of everything like an animal presence, rusted and spackled with Bondo, serene and almost majestic. A cat with luminous green eyes observes me from the perfect darkness between a matching washer and dryer. I go up the back stairs. The stairs have been carpeted with individual strips of green shag; the strips have been fastened down with bright red and yellow thumbtacks. I unlock the back door with a paperclip. Then I'm standing in the redecorated kitchen. Everything gleams in the darkness, aided by moonlight which falls through the cafe-style curtains. Nothing looks familiar here either, and I move into the living room.

We learn the rules when we get older, and that's what helps us get by. We're not uncertain anymore. We're not startled by the slightest sounds. As I step into the living room the only thing I find familiar here are the floorboards, which do not creak when I don't want them to. I feel like a spider on its home web, exerting texture, balance and pressure, gliding across the surface of spaces and silver fabric. Small children never know. They don't know why people do things, or even what they're going to do next. Small children invent their own reasons for things and things that happen. Children are reasonable too, just like adults. It's just that children don't know the acceptable rules of reason yet. Children can get lost. They need someone strong to lead them. Otherwise, they can be easily led astray by the convolutions of their own minds. Childhood is not a glorious thing. Childhood does not comfort or instruct. Childhood isolates people. Sometimes, children make mistakes which they regret later on in their lives.

This living room was not filled with fine furniture, but it was clean and functional and what is often referred to as homey. Flowers in vases, framed photographs on shelves, a crushed velvet family portrait of the Kennedys, dull paisley wallpaper, a large waiting television console, the whirling dust and fading, sun-bleached curtains. Only this linked me with the past, this whirling dust. This was the vast sound into which Mom had vanished. I thought I heard something and I turned. A tiny rectangle of light escaped from underneath one of the bedroom doors.

"The history of motion," I whispered. "The history of motion. Light, sound, heat, gravity, mass, life. Motion, the history of motion," I whispered, over and over again, but I couldn't remember, the fundamental weight of the words didn't work. "The history, the history of light. Light and history, history and motion, motion and history . . ." The words were like the dust. They whirled without the frame of sentences or that dark ritual meaning which would call her forth again, out of the shadows of this lost house. The words would make the house ours again. The words would bring Mom back to earth. "The history of motion. The history of motion, Mom . . ." My family was very far away and inaccessible to me now. But I didn't want to be with them. I wanted to be here. I wanted to stay here forever. I couldn't stop crying, but I didn't have to stop crying either. It wasn't that I wasn't happy. It wasn't that I didn't know my life had turned out for the best. But I was growing up now, and so I could cry all I wanted to, even while my hands filled with the tears, even while I felt the icy wet turning in my stomach and my heart. "The history, the history of motion." And then, as if by magic, the door in the hallway opened. A mist of pale yellow light poured out.

The child rubbed its eyes with its fists and looked at me.

"*¿Quien es?*" she asked. She was a dark formless shape, looking for the bathroom. I must have looked enormous to her.

She pulled her tiny fists away from her dark eyes, peering at me, incredibly tiny and perfect like a tiny robot. "*¿Eres tu mi padre? ¿Eres tu?*"

This was it. I was about to find what I could not find in my own big house with all the strong, independent and well-designed furniture I possessed there. If only I could remember, I could stay. I wouldn't have to go back. Everything would be better again, it would all make perfect sense: the history, the history of motion, something about light, and living one's own life in a world where everything is always moving, and the way time takes you away from people, even people you love. . . . But I couldn't remember how it went.

I couldn't remember the tone or the light of it, the chords or the melody. . . . The words wouldn't come and I couldn't make them.

We stood looking at one another in the dark hallway. She was not that much smaller than me. In some ways, however, she was very much smaller.

"Mi madre esta en la cama. La cama de mi madre es azul."

There was no way back unless I could remember. And I couldn't. I couldn't remember. I couldn't remember.

"Carmelita?" a woman's voice said. "Carmelita? *¿Qué pasa, pobrecita?* What are you doing?" The woman's voice was growing larger and more indistinct.

I continued to cry, without any reason. I wasn't even sad. I didn't feel lost or lonely or forlorn. I just stood there in the heavy darkness and saw the light click on in the open doorway down the hall. I heard slippers being pulled onto feet, then the feet on the floor. In a moment, the child's mother would appear at the bedroom door.

ABOUT THE AUTHOR

Scott Bradfield was born in California in 1955. He taught
for five years at the University of California, Irvine,
where he received his doctorate in American literature,
and presently teaches English at the University of Connecticut at Storrs.
He is also the author of the short-story collection
Dream of the Wolf.